Population Harvesting

MONOGRAPHS IN POPULATION BIOLOGY

EDITED BY ROBERT M. MAY

(list continues following the Index)

Population Harvesting

Demographic Models of Fish, Forest, and Animal Resources

WAYNE M. GETZ

AND

ROBERT G. HAIGHT

PRINCETON UNIVERSITY PRESS

PRINCETON, NEW JERSEY

1989

Published by Princeton University Press,
41 William Street, Princeton, New Jersey 08540
In the United Kingdom: Princeton University Press, Guildford, Surrey

Library of Congress Cataloging-in-Publication Data
Getz, Wayne M., 1950–
    Population harvesting: demographic models of fish, forest, and
animal resources.
    Wayne M. Getz and Robert G. Haight.
        p.   cm.—(Monographs in population biology; 27)
    Bibliography: p.
    Includes index.
    ISBN 0-691-08515-3 (alk. paper)
    ISBN 0-691-08516-1 (pbk.)
    1. Population biology—Mathematical models. 2. Wildlife
management—Mathematical models. 3. Forest management—
Mathematical models.
    I. Haight, Robert G., 1955–   . II. Title. III. Series.
    QH352.G47 1989
    574.5'248—dc19   88-19950

This book has been composed with the T_EX system in Autologic
Baskerville II by Zoological Data Processing, Socorro, New Mexico,
and set at T_EXSource, Houston, Texas

Clothbound editions of Princeton University Press books are printed
on acid-free paper, and binding materials are chosen for strength and
durability. Paperbacks, although satisfactory for personal collections,
are not usually suitable for library rebinding.

Printed in the United States of America by Princeton University Press,
Princeton, New Jersey

To our wives

Jennifer and Georgiana

Contents

CONTENTS

CONTENTS

CONTENTS

Preface

Although the theory of harvesting populations has a number of different traditions drawn from fisheries, forest, and wildlife management, a common demographic thread runs through these various applications. Trees and many vertebrates reproduce on a seasonal basis so that their populations consist of cohorts of similarly aged individuals (age classes). Thus discrete age- and, more generally, stage-structured (e.g., size classes) population equations are appropriate for modeling the dynamical aspects of both animal and plant populations.

One of the aims of this book is to draw together the theory of discrete stage-structured population models as developed in the fisheries, forestry, population harvesting and general demography literature. We do this in the specific context of biological resource management. The disciplines of fisheries, forest-stand, pest, and wildlife management have their own unique problems, but common economic and demographic notions pervade the mathematical analyses of these problems. We hope, by unifying some of these notions across the various areas of application, that this book will encourage a cross-fertilization of ideas between professional fisheries, forest, pest, and wildlife management scientists, as well as population biologists and demographers.

A second aim of this book is to present a comprehensive account of our recent investigations into the theory of nonlinear stage-structured population harvesting models and its application to fisheries and forest-stand management problems. The linear theory of age-structured population growth is embodied in life-table analysis (static viewpoint) and Leslie matrix theory (dynamic viewpoint). Nonlinearities, however, are an essential aspect of biological systems, the most obvious being increases in mortality and reduction in fecundity rates as population density increases in a resource-limited environment. Because a general linear theory is sufficiently extensive to warrant a book on its own, we only summarize this theory in Chapter 2 and provide the material necessary to achieve continuity with the applications presented in Chapters 4 to 6 and the nonlinear theory presented in Chapter 3.

Most of the advanced material presented here appears or will appear in the recent literature and references are provided, although a small percentage of the material is not published elsewhere. Our treatment assumes that the reader is comfortable with basic notions in calculus, matrix algebra, and complex number theory. Discrete models allow us to avoid some of the more difficult aspects of mathematical analysis associated with systems of differential and integro-differential equations. As this is an advanced rather than an introductory text, we assume that the reader is familiar with the basic elements of matrix algebra and complex numbers. We do lead the reader through a cursory treatment of matrix diagonalization (eigenvectors and eigenvalues) and the solution to linear matrix equations, but expect those readers who have difficulty with the concepts to supplement their reading using the references provided. We cover some aspects of linear and nonlinear programming, including a discrete version of Pontryagin's Maximum Principle, but only the minimum necessary to provide a self-contained presentation of the material in this book.

The material in this book should be accessible to those forest and fisheries economists and modelers who have read such books as Clark (1976) or Johansson and Löfgren (1985). We hope, however, that this book will be of value to population and wildlife biologists who only have an elementary background in calculus and matrix algebra, but are motivated to work hard and insert supplementary

PREFACE

readings when the going gets rough. In particular, these
readers can omit the more difficult sections, 2.4 and 3.3 to
3.6, and still follow and appreciate much of the material
presented in Chapters 4 to 6. We also hope that this work
motivates applied mathematicians interested in resource
management and/or population harvesting to study some
of the more general properties of discrete nonlinear stage-
structured models.

The ideas in this book draw strongly from our collabo-
rations with colleagues. In particular, W.M.G. is indebted
to R. C. Francis and G. L. Swartzman for many stimulat-
ing discussions over the past eight years while working on
joint fisheries projects supported by the Northwest and
Alaska Fisheries Center of the National Marine Fisheries
Service. R.G.H. is indebted to D. Brodie and D. W. Hann
for supervising his dissertation research which led directly
to the forest management studies described in Chapter 5.

We typed this manuscript using LaTeX, a macro pack-
age written by L. Lamport for the TeX typesetting system
developed by D. E. Knuth. To these two people, we are
indeed grateful. We are also indebted to M. E. Bergh,
L. W. Botsford, D. Goodman, J. E. Hightower, J. Schnute,
B. Solberg, G. L. Swartzman, and D. Wear for critically
reviewing drafts or portions of this manuscript. Of course
all errors and omissions remain our own responsibility. We
thank Nancy Munro for drafting the figures.

We would like to thank Judith and Robert May, our ed-
itors at Princeton, for their support and help in getting
this book published.

W.M.G. would like to acknowledge the National Science
Foundation (Grant DMS-8511717) and the Alfred P. Sloan
Foundation (Grant No. 86-6-18) for support towards the
research and writing of this book. R.G.H. would like to
acknowledge support of an S. V. Ciriacy-Wantrup Postdoc-
toral Fellowship administered by the Graduate Division at
the University of California, Berkeley.

Population Harvesting

CHAPTER ONE

Introduction

1.1 SCOPE

1.1.1 Preview

This book is about resources that are managed by harvesting cohort-structured biological populations. This includes fisheries, forest-stand, and wildlife management problems, as well as mass rearing of insects for biological control. Our aim in writing this book is to present a unified approach to modeling and managing such resource systems.

In this age of environmental crises we need to meet the challenge of managing biological resources in an efficient and minimally disruptive manner. This requires that we be precise about what we are doing; and to be precise we need to model the management process. In mathematical terms, efficient management typically translates into maximizing some suitably defined performance index, often net revenue. Precision requires that we model the dynamic response of the underlying population to management actions.

Population modeling is an inexact science. Populations are part of complex systems that defy the taming tethers of "physical laws." Thus there is little to prevent a piecemeal approach to cohort-structured resource management, with each subdiscipline developing its own methods. To a large extent this has happened, and communication among scientists working on conceptually similar problems in different areas of applications has been hindered. Communication among scientists working on managing cohort-structured populations, albeit as different as trees, fish, mammals, and insects, can only be beneficial.

3

We hope this book will encourage communication among scientists working in different areas of application.

The first requirement in solving the communication problem is to develop a common language, especially in the context of population modeling, where the greatest barrier exists. We do this by adopting a neutral mathematical notation: the notation of mathematical systems theory. This has the added advantage of using a notation that is more suitable for the mathematical analysis of management problems than notations that are currently often used in the applied fields.

The second requirement in solving the communication problem is to develop a modeling framework that can be applied to any cohort-structured resource management problem. This facilitates comparative analyses of conceptually similar problems in the different areas of application, preventing duplication of effort and enhancing our general understanding of resource management issues. In the theory sections of this book, we develop an approach that allows us to incorporate such nonlinearities as density-dependent reproduction and survival, while retaining most of the clarity associated with linear population models. Such clarity is not apparent in current non-linear approaches.

The various areas of resource management remain distinct not only because of differences in the biological species comprising the resource, but because each area of application poses a different set of problems. A major emphasis in fisheries science has been on the problem of estimating current and past population levels (i.e., stock abundance) using catch levels and fishing effort data (see Cushing, 1981; Gulland, 1983; Schnute, 1985). Many fish populations, especially those in which individuals live no more than several years, exhibit wide and largely unpredictable fluctuations in the number of young fish (i.e., recruits) joining that part of the stock that is vulnerable to fishing each year. Thus stochasticity is a critical aspect of analyzing the stock dynamics in most fisheries. This stochasticity, primarily due to environmental changes and the problems associated with estimating population abundance and age structure, poses severe constraints on our ability to develop appropriate management policies. These difficulties have led to a dichotomy in methodology, namely "cohort" and "surplus production" approaches to yield or catch analysis.

Origination of the "cohort" approach is largely due to Beverton and Holt (1957), who developed a method of analysis in which the age of the fish play a central role. Beverton and Holt's approach was essentially a deterministic equilibrium analysis which assumed constant recruitment. This approach has been extended to include nonlinear recruitment (Getz, 1980a,b; Reed, 1980), and dynamic (Getz, 1985, 1988) and stochastic (Reed, 1983; Getz, 1984a) analyses, but the multidimensional character of the models (model variables are age classes) makes the analysis complicated.

The "surplus production" approach typically ignores age by focusing on a single harvestable stock biomass variable (single variable models are sometimes referred to as "lumped-variable" models). The analysis leads to the derivation of a scalar catch equation (Baranov, 1925) that is more readily embedded into a nonlinear stochastic setting (Schnute, 1985). Although cohort structure is essentially ignored, the value of this approach lies in being able to analyze highly stochastic management situations (see Walters, 1986). There are some important drawbacks, however, to ignoring cohort structure when undertaking a detailed stochastic analysis. The market value of individual fish and our ability to catch them may vary quite considerably with age (or its correlate, size). Although we take a cohort approach throughout this book, in the fisheries chapter (Chapter 4) we

5

will demonstrate a link between cohort-structured and lumped-variable models.

In contrast to fisheries, individuals in a forest stand can be seen and counted, and tree attributes such as stem diameter, height, and volume are readily measured. In addition, the time units associated with birth, growth, and death processes for trees are at least an order of magnitude larger than for fish populations. Thus, density-dependent models for projecting stands of trees are based on short-term growth observations in a cross section of stands with a wide range of ages, densities, and environmental conditions.

The relative ease of measuring and harvesting trees has resulted in a variety of stand management systems and associated models for predicting growth and determining harvest policy. Models for projecting stand growth span the range from single-variable models for single-species even-aged stands to single-tree simulators for mixed-species uneven-aged stands. The stage-class model that we apply to forest stand projection has an intermediate level of biological detail and complexity. The stage-class model includes the important demographic processes such as species- and size-dependent birth, growth, and death processes. As a result, the same stage-class model can be used to project the effects of harvesting in either even-aged or uneven-aged stands with one or many species. Equally important, the stage-class model is simple enough for mathematical analysis and numerical optimization. Thus, the economic efficiencies of different timber harvesting systems can be evaluated and compared.

The utility of the stage-class model in forestry research is illustrated by the development of the theory associated with uneven-aged stand management. Prior to the application of stage-class models in forestry, uneven-aged management problems were formulated only in a steady-state framework because of the limitations of growth and yield models in use. Univariate models did not give realistic portrayals of stand dynamics in uneven-aged stands, and single-tree simulators were too complicated to analyze in a dynamic framework. The steady-state analyses often concluded that clearcutting and plantation management were the most efficient management system. In contrast to univariate models and single-tree simulators, the stage-class model provided a means to formulate and solve a dynamic harvesting problem. Results from the application of the dynamic model, as discussed in Chapter 5, show that clearcutting and even-aged management are just a special case of the more generally formulated uneven-aged management problem and are less efficient in the long term.

The problem of counting and measuring large-mammal populations (both terrestrial and marine[1]) lies somewhere between fish and tree populations. One can more easily count mammals than fish and one can more easily estimate the size of individual mammals than trees. Of course, different mammal populations present different problems. Pinnipeds (seals, sea lions, walruses), for example, are easily observed only during their breeding season when they congregate in specific areas.

Insects present an altogether different challenge. Many species have multiple generations per year, where the length of each generation is strongly linked to the ambient temperature of its habitat. Survival rates fluctuate dramatically with environmental conditions so that populations in the field can exhibit dramatic changes from week to week. The methods of analysis described in this book are largely unsuitable for addressing insect pest management problems, although we illustrate how they can be applied to insect populations reared under controlled laboratory conditions.

---

[1] Exploitation of a marine mammal population is traditionally referred to as a fishery, although we treat the problem separately in this book.

As previously stated, the common thread running through managing fisheries and forest stands, exploiting large mammals, and mass rearing insects is the harvesting of populations that have a cohort structure. In analyzing the growth of these fish, tree, mammal, and insect populations, we usually categorize individuals by age or size and collect data to construct an age- or size-specific life table. One limitation with the age-specific approach, however, is that it is not always possible to determine the correct age of an individual. Thus size may often be a more appropriate variable.

In forest-stand management, in particular, individuals are assigned to size or "stage classes" rather than age classes. A size-specific analogue of a life table can be constructed for such populations, but this introduces other complications such as assessing the rate at which individuals move from one size class to the next. The application of age-specific life tables to certain insect populations also presents special problems. Holometabolous insects, such as flies, moths, butterflies, beetles, bees, ants, and wasps, have distinct life stages (egg, larval, pupal, adult) with, for example, very different feeding behaviors in each life stage. Because not all individuals spend the same period of time in each life stage, a more general age-stage life table is required.

In this book we focus on age- and stage-structured approaches to modeling populations and their application to problems in fisheries and forest-stand management, in harvesting large mammals, and in mass rearing insects. Although we do not deal with lumped population models (i.e., models in which all age classes are lumped together), we show how to collapse age-structured models to lumped-variable stock-biomass models that relate stock levels at several points in the past to current stock levels. These models are so-called discrete-delay difference equations. Since they capture certain essential features of age-structured models, some of their theoretical properties and their application to fisheries problems are discussed in this book.

Most resources are derived from populations that exhibit cyclic reproduction. For these populations, systems of discrete-time difference equations provide a more appropriate model of the resource than systems of continuous time differential equations. Compared with differential equations, difference equations are easier for the nonmathematician to construct and simulate on the computer. Thus we will not attempt to extend any of the theory presented here to continuous system models.

1.1.2 *Focus*

The field of biological resource management draws its tools of analysis from mathematics, statistics, computer science, and engineering. Thinking in this field is influenced by economists and other social scientists. Economically important resource systems include both plants and animals, in both marine and terrestrial environments.

The various resource management professions (fisheries, forestry, and wildlife management) are distinct from one another. Resource economists and modelers, however, provide a bridge between these professions (see Clark, 1976). The same mathematical and statistical techniques are often employed in different areas of applications (e.g., mathematical programming), although each area has its own distinct flavor. For example, multidimensional models and linear and nonlinear programming techniques are used extensively in forest management, while dynamic scalar models and statistical estimation theory are important tools in fisheries management. Thus resource managers in different areas of application have a different quantitative training; and this tends to keep managers, and even scientists, working in these different areas further apart than they should be. Additionally, even within specific areas of application, some researchers are concerned primarily with the statistical problems associated with estimating model parameters, while other researchers are concerned with determining harvest rates and the dynamic response of the population to management.

The material in most resource management books is a vertical slice through the field in that it deals with a single area of application such as fisheries (e.g., Cushing, 1981; Gulland, 1983; Rothschild, 1986) or forests (e.g., Husch *et al.*, 1982; Johansson and Löfgren, 1985). Examples of books taking a horizontal slice are Clark's development of bioeconomic principles (Clark, 1976), Mangel's synthesis of uncertain resource systems management (Mangel, 1985), and Walters' exposition on the adaptive management of resource systems (Walters, 1986). This book is horizontal since its general theme is harvesting of age- and stage-structured populations. Our models are nonlinear extensions of discrete-time, age- or stage-class matrix models that can be traced back to Lewis (1942) and Leslie (1945, 1948). In Chapters 2 and 3 respectively we develop linear and nonlinear mathematical techniques of analysis and derive a small number of theoretical results. In Chapters 4 and 5 respectively we develop models for application in fisheries and forest-stand management, and include a number of literature case studies to illustrate how various types of resource management problems are approached. In Chapter 6 we illustrate how age- and stage-structured models have been applied to a few selected animal resource and pest management problems.

We have limited the scope of the book to dealing with questions that relate only to harvesting, and then only in populations where age or stage structure is regarded as an essential component of the management problem. This was done to keep the book a manageable size, but as a result we do not present any of the theory relating to resource models in which populations are represented by a scalar population-biomass or population-numbers variable (i.e., lumped-variable models). For a treatment of this theory the reader is referred to Clark (1976, 1985). We also do not deal with the extremely important issue of how to estimate model parameters. Such questions rely heavily on both statistical analyses and the biology of the populations involved. Many of these questions are more appropriately dealt with in an application-specific vertical text. Further, the difficult statistical questions are best treated in a book that includes sufficient statistical background material to assist the reader with the development of the more sophisticated techniques. Here we only provide background material in discrete dynamic systems and age- and stage-structured models. We do include techniques for modeling and analyzing stochastic problems, but our stochastic treatment is limited to modeling questions. Stochastic analyses of structured systems present a major challenge, although some progress has been made (for example, see Horwood, 1983; Horwood and Shepherd, 1981; Reed, 1983; Nisbet and Gurney, 1982). A deep understanding of the influence of stochastic processes on harvesting is difficult to obtain in multidimensional models. Most of our understanding stems from simulation studies. The influence of stochastic processes is better understood in the framework of scalar or lumped-variable models (for example, see Beddington and May, 1977; May *et al.*, 1978; Ludwig and Walters, 1981; Walters and Ludwig, 1981), although this material is not presented here because the corresponding deterministic models are outside the scope of this book.

A central issue in applying models to particular problems is deciding how much biological detail should be included in a model. The models presented in this book include age or stage structure in the population, but ignore specific characterization of interactions between the

11

population and its predators, competitors, prey, or physical environment. Is this a reasonable approach? This is a difficult question to answer. We address this question below, in a discussion that is intended for readers that already have some modeling experience. Other readers should move on to Section 1.2.

1.1.3 Model Resolution

There are two fundamentally different sets of paradigms associated with modeling the dynamics of biological populations: the *top-down* and *bottom-up* approaches. In our discussion of these modeling paradigms we refer to *precision* and *accuracy*, where a lack of precision implies a large variance associated with model predictions and a lack of accuracy implies a bias in model predictions. A key question in modeling is when does increased biological detail lead to improved precision and/or improved accuracy? Intuitively it would seem that by incorporating additional biological mechanisms we can account for a greater variance in the population's behavior and, hence, construct more precise models. On the other hand, if some details are incorporated but a major component of the model is missing, a detailed model may be quite inaccurate when compared with a less detailed but intrinsically correct model.

To be more specific, suppose $x(t)$ represents the number of individuals in a population at time t, and let $b(t)$ and $d(t)$ respectively denote the number of births and deaths in the population over the time period $[t, t+1]$. Then the number of individuals at time $t+1$ is given by the equation

$$x(t+1) = x(t) + b(t) - d(t). \qquad (1.1)$$

The top-down approach assigns some functional form to the birth and death processes $b(t)$ and $d(t)$, such as

$$d(t) = \bar{d}x(t), \qquad (1.2)$$

where $\bar{d}$ is a constant per capita death rate. The bottom-up approach, on the other hand, identifies $d(t)$ as the sum of all deaths stemming from a number of different sources, including losses to possibly a number of different predators, for example, as well as death from a number of different diseases, etc. If an important source of mortality is omitted in the latter approach, however, then the best holistic estimate of the overall death rate $\bar{d}$ in relationship (1.2)[2] may provide a more accurate model than the bottom-up approach.

In general, the bottom-up modeling paradigm focuses on detail, is problem specific, and, in a sense, is an empirical approach; that is, modeling progresses by using biomass or energy flow equations to link those and only those components that are observed to be part of the system (Forrester, 1961; Odum, 1983; Swartzman and Kaluzny, 1987). In the bottom-up approach we explicitly recognize that populations are part of a complex ecosystem and we incorporate underlying ecosystem processes into a dynamic model of one or more populations of interacting individuals. Models are developed much in the same way as one develops computer software to solve specific problems. So-called ecosystem compartmental simulation models are an example of this approach (for specific examples, see Patten, 1971, 1972). In bottom-up modeling one generally opts for realism through detail, but at the same time one endeavors to simplify a model without sacrificing accuracy or substantially affecting precision.

In the top-down modeling paradigm we tend to ignore detail. This approach is more applicable than the bottom-up approach to addressing a general class of problems and, in a sense, is a phenomenological approach; that is, modeling begins by postulating a fundamental growth

---

[2] This estimate is obtained by counting the number of dead in each time period, or inferring the number that died from past and present population levels.

law for the population. (Equation (1.2) is a poor example of such a law. For better examples see Getz, 1984b, and Ginzburg, 1986.) In the top-down approach we recognize that populations are part of a hierarchical system with ever-increasing levels of complexity. We begin modeling at the simplest of these levels, and usually do not move beyond one or two additional levels of complexity. Models are holistic: model building begins by postulating a basic growth law (Getz, 1984b), employing a conservation equation (Schnute, 1985, 1987), or describing the demographic processes of mortality and natality in terms of average rates that summarize the cumulative effects of underlying but unspecified ecological factors. In top-down modeling one typically explores that trade-off between increased detail (which often results in a loss of comprehension) and development of principles (which we hope enhances our understanding). Certainly, one should not add details that do not increase the accuracy or precision of the model.

In population modeling, there is a place for both paradigms. The particular approach taken depends on the objectives of the modeling study. The bottom-up approach provides a method for structuring and coordinating ecosystem studies. The top-down approach provides a method for exploring theoretical questions, for example, relating to the effects of intra- and interspecific competition on population growth and stability (Getz, 1984b). It is one thing to simulate the response of a population to a number of environmental inputs and another to solve for optimal harvesting or management strategies. The latter requires extensive numerical work that is facilitated by simple models described by mathematical functions that are, for example, differentiable. Thus top-down models are going to be more easily applied to resource management problems. If we are able to show, as we do in the forestry chapter (Chapter 5), that a top-down model has essentially the same precision as a bottom-up model in predicting management-related variables, such as biomass yield, then the top-down model is preferable for the management study.

Because of the management orientation of this book, we take a top-down approach throughout. In addition to the greater generality and ease of implementation of top-down models, the amount of work required to fill in the biological details of a bottom-up approach would detract from research on management-oriented questions. Age and/or stage progression of individuals is included in our population models because such progression is a general phenomenon that has economic and logistical (implementation of strategies) implications for management. Details of ecological interactions tend to be case specific. Thus in the top-down approach it is appropriate to subsume these interactions in the form of density-dependent relationships and stochastic inputs, and retain some generality.

This is especially true of resource management studies where we focus on finding harvesting strategies that satisfy certain performance criteria rather than on explaining basic biological processes. In forest-stand management, for example, the models are not constructed to investigate the physiology of tree growth for which a bottom-up approach may be necessary. Rather, they are designed to assess the response of a stand of trees to various harvesting practices, and provide a tool for finding "optimal" harvesting strategies. Suppose a detailed physiological model were available that could accurately predict the growth of a tree as a function of its environment. Since every tree grows under a different set of local conditions ranging from soil type and water runoff rates to level of competition from immediate neighbors, a model that includes nutrient uptake and competition at the individual-tree level is impossible to implement at the stand level without investing a great deal of time and money just to set up the initial conditions to begin the simulation. Thus some compromises are required, such as using data that reflect average growth and survival rates for classes of trees. The accuracy involved in estimating these averages depends on both the quantity and quality of data. There is an obvious marginal value associated with the collection of such data. Certainly, a manager would not want to invest the value of the stand in collecting data to determine the optimal harvesting regime. Our experience suggests that there is a negative marginal utility to assessing more than an average age- or stage-specific growth and survival parameter plus a couple of regeneration or reproductive parameters. The reason for this, as we discuss in Chapter 5, is that the relatively simple stage-structured models presented in this book seem to be indistinguishable from more detailed single-tree stand-growth simulators in predicting stand yield, at least at the level of accuracy determined by the quality of the data used to verify these models.

In fisheries, the limitations on biological detail are even more severe. The high degree of variability associated with survival of youngest age class (that is, individuals in the egg and larval stages) plus the difficulties associated with estimating the size of a population or even the age of individuals taken from the population limit the extent to which detailed age structure can be included in models. Nevertheless, as previously mentioned, the size of harvested individuals often has economic ramifications that cannot be ignored. For this reason, models that have age structure may prove to be more valuable than lumped models for making certain types of management decisions. It should be pointed out, however, that fisheries models which directly use time series of catch-effort data, say, to predict future population or yield levels without independently assessing model parameter values from biological data (e.g., Walters, 1986) are often more robust if age or stage structure is ignored. Time-series models are central to the analysis of many resource problems (Schnute, 1985, 1987), but are beyond the scope of this book.

1.2   HOW TO READ THIS BOOK

Since readers of this book will range from quantitative problem solvers (mathematicians, physicists, engineers) interested in resource management problems to population biologists interested in a particular resource application, we feel that it would be useful to have a guide to reading this book. A number of sections were not written for the general reader. In particular, Section 3.4 is only included for the more quantitatively oriented reader, especially those interested in dynamic optimization. Also, most of the material in Section 3.5 can be skipped over by those readers interested in Monte Carlo techniques for simulating stochastic phenomena, rather than directly modeling the time evolution of the population distribution or its first two moments.

The following is a list of selected sections that should be read by those with one of the identified interests. The reader interested in fisheries or forest-stand management should begin, respectively, with a preliminary reading of Chapter 4 or 5 before earnestly tackling the theory presented in Chapters 2 and 3. This will give that reader a feel for where the theory is heading in terms of particular applications.

- **Deterministic harvesting theory**: Sections 2.1–2.3; 3.1, 3.2, 3.3*, 3.6; 4.1–4.4, 4.6; (5.1–5.3)* (sections with an asterisk may be omitted by those readers not interested in stage-class models).
- **Stochastic harvesting theory**: To the above sections add Sections 2.4; 3.5; 4.5.

- **Fisheries management**: Skim Chapter 4 and then read Sections 2.1, 2.2.1–2.2.2; 3.1, 3.2, 3.5.6, 3.6. Reread Chapter 4.
- **Forest-stand management**: Skim Chapter 5 and then read Sections 2.1–2.3; 3.1, 3.3. Reread Chapter 5.
- **Animal demography**: Chapter 2; Sections 3.1, 3.2; 4.1–4.3; 6.1, 6.2.

The reader should note that theorems, lemmas, and results are numbered sequentially in the chapter in which they appear, with reference given to chapter number (e.g., Theorem 2.1, Lemma 2.2, Result 2.3). We also present five problems that are numbered sequentially throughout the book without reference to the chapter number in which they appear (e.g., Problem 1, Problem 2, etc.).

Linear Models

2.1 LIFE TABLES

2.1.1 Scope

Life-table analysis uses the average fecundity and mortality schedule of individuals within a population to study the growth and cohort or age-structure dynamics of that population. Classical life tables are constructed under a number of severe assumptions. Fecundity and mortality processes are assumed to be unaffected by population density and to be time invariant (stationary with respect to time). The reproductive success of females is assumed to be unaffected by the number of males in the population, and all individuals are subjected to the same fecundity and mortality schedules. Let the continuous variable a be used to denote age. All females in the population are assumed to be subject to the same

- birth processes, characterized by an age-dependent female fecundity (natality) rate or *force of maternity* function ϕ_a;
- death processes, characterized by an age-dependent female mortality rate or *force of mortality* function μ_a.

These assumptions can be used to construct a life table for the female component of the population, or for the population as a whole if a constant sex ratio is assumed.

In real populations we know that mortality and fecundity rates change with time, depending on environmental conditions, population density, and the sex ratio in the population. Insofar as population density and sex ratio can be assumed to play a minor role in determining mortality and fecundity schedules, and as time or seasonally related variations in parameter values around average values can be ignored, a deterministic theory of population growth and age structure can be developed using the functions ϕ_a and μ_a defined above. Since we focus on populations that live for at least several years and exhibit an annual breeding cycle, it is convenient to select one year (for animals which live on the order of tens of years) or five years (for trees which live on the order of hundreds of years) as an appropriate unit of time, and to discretize the analysis using an integer index $i = 1, \ldots, n$, where n is the oldest cohort (age category) considered in the analysis. Of course, other time units have been used, including 2 years for such long-lived animals as elephants and whales. In the case of insects, 1 or 2 days is the time unit commonly used.

In this chapter, it is our intention to review only that part of life-table analysis that relates to the development in Chapter 3 of a nonlinear demographic theory for the management of animals and timber. Other texts are available that treat life-table analysis in the context of classical (human) demography (Henry, 1976; Keyfitz, 1968; Pollard, 1973), and evolutionary biology (Charlesworth, 1980).

2.1.2 Mortality Column

Suppose that l_0 individuals are born at time $t = 0$ (typically l_0 is taken as 1000 for the presentation of life table mortality data). Then, since the force of mortality is assumed to be dependent on an individual's age but is independent of population density and time—that is, $\mu_a(t) = \mu_a$ is the per capita mortality rate for all values of t—the number of individuals l_a in this cohort that live to time (age) a by definition satisfies the differential equation

$$\frac{dl_a}{dt} = -\mu_a l_a,$$

which can be solved to yield at the integer time $t = i$

$$l_i = l_0 e^{-\int_0^i \mu_a \, da}.$$

Since the age-specific mortality rates are assumed to be stationary with respect to time, it follows that l_i/l_0 is the proportion of individuals that survives up to *exact* age i, irrespective of whether these individuals were born at time 0 or some arbitrary time t. To simplify notation, we will assume throughout that the mortality schedule l_i has been normalized to directly represent proportions; that is, $l_0 = 1$. It also follows that s_i, the proportion of individuals (alive at age i) that survives from exact age i to exact age $i+1$ is independent of time t and is given by

$$s_i = \frac{l_{i+1}}{l_i}, \qquad i = 0, 1, 2, \ldots . \qquad (2.1)$$

Note that the positivity and monotonicity of l_i imply that $0 < s_i \leq 1$, for all i except the oldest age group. For the latter, by definition, we have $s_n = 0$. Also, equation (2.1) leads to the relationship (recalling that $l_0 = 1$)

$$l_{i+1} = \prod_{j=0}^{i} s_j. \qquad (2.2)$$

The values $l_0, l_1, \ldots, l_n$ listed vertically in a life table are referred to as the *survivorship* column. (For an example see Table 6.1). From this column a number of other life-table entries are generated, including

$$q_i = (l_i - l_{i+1})/l_i,$$

the proportion of individuals that die between exact age i and exact age $i+1$, and (see Emlen 1984 for a derivation)

$$e_i = \left(\sum_{j=i+1}^{n} l_j \right) \Big/ l_i,$$

the life expectancy of an individual at exact age i (note that in the formula for e_i, n must be the oldest age beyond which no individuals are known to survive). In continuous time, the values of $\log l_a$ plotted as a function of

21

a are referred to as the survivorship curve for the population in question. Three classes of survivorship curves are distinguished (Figure 2.1), respectively typified by very low infant and juvenile mortality rates (e.g., man), constant mortality rates with respect to age (e.g., some birds), and very high infant mortality rates (marine invertebrates, many species of fish).

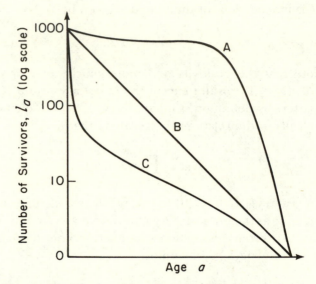

FIGURE 2.1. The logarithm of survivorship curves l_a for three contrasting life histories: A, low infant and juvenile mortality rate; B, constant mortality rate with age; C, high infant mortality rate.

2.1.3 Reproductive Cycle

Subject to the biological problem of determining the age of each individual in a population, the population can be divided into age classes. If populations have a regular reproductive cycle (usually annual), we can count the population immediately after births have taken place and set the time axis so that births occur at integer values of t (or integer multiples for long-lived species). If this occurs at time t, then the variable $x_i(t)$ can be used to denote the number in the ith class, $i = 0, \ldots, n$, at time t, where $x_0(t)$ is the number of newborns. Note that n is either the maximum age to which any individual can live or, more generally, n represents all the individuals that have just turned n years and older.

It now follows from our definition of the survival parameters s_i in equation (2.1) that, over the reproductive cycle $[t, t+1]$, $t = 0, 1, 2, \ldots$, variables x_i will satisfy the relationship

$$x_{i+1}(t+1) = s_i x_i(t), \qquad i = 0, \ldots, n-2, \qquad (2.3)$$

with this same equation holding for $i = n-1$ if $s_n = 0$, that is, if no individuals survive beyond age n. If $s_n > 0$, and it is assumed that s_n applies to all individuals older than n, then the equation for x_n becomes

$$x_n(t+1) = s_n x_n(t) + s_{n-1} x_{n-1}(t). \qquad (2.4)$$

To complete our description of the population projection process from time t to $t+1$, we need to consider the production of newborns x_0 at time $t+1$. In a continuous model, we would need to integrate the natality function ϕ_a with respect to the female age distribution to obtain the actual rate of births at time t. However, if we assume that females only reproduce on their birthdays, so to speak (which is a good approximation if the breeding season is short with respect to the breeding cycle), then the natality function ϕ_a is replaced by a set of constants b_i denoting the expected number of females per aged i female (recall that we are either modeling a population with a constant sex ratio or we are modeling the female portion of the population under the assumption that there

23

are always sufficient males to allow normal levels of reproduction). In this case, the number of newborns is given
by the following sum:

$$x_0(t + 1) = \sum_{i=1}^{n} b_i x_i(t + 1).\qquad(2.5)$$

The system of equations (2.3)–(2.5) thus allows us to
calculate the number of newborns $x_0(t + 1)$ and the pop-
ulation vector $\mathbf{x}(t + 1) = \big(x_1(t + 1), \ldots, x_n(t + 1)\big)'$ (that is,
the number of individuals in each cohort) at time $t + 1$, if
we are given the survivorship and fecundity parameters s_i
and b_i for all ages i and the population vector $\mathbf{x}(t)$ at time
t. (Note throughout the text that $'$ is used to denote the
transpose of a vector, since the standard representation of
a vector is a column rather than a row of elements.) Alternatively, if the number of individuals x_0 born at times
$t, t - 1, \ldots, t - n + 1$ is known, then it follows from the iterative application of equation (2.3) and relationship (2.2)
that (recalling $l_0 = 1$)

$$x_{i+1}(t + 1) = l_{i+1} x_0(t - i), \quad i = 0, \ldots, n - 2,\qquad(2.6)$$

which is another way of projecting $x_0(t + 1)$ and $\mathbf{x}(t + 1)$.
Note that if equation (2.4) holds with $s_n > 0$, it is not
possible to express $x_n(t+1)$ in terms of $x_0(t-i)$ (unless $x_0(t)$
is assumed to be constant as in the case of some fisheries
models), since $x_n(t+1)$ in fact depends on $x_0(t-i-1), x_0(t-
i - 2), \ldots$, etc. On the other hand, if $s_n = 0$ in (2.4) then
equation (2.4) can be written as

$$x_0(t + 1) = \sum_{i=1}^{n} b_i l_i x_0(t - i + 1),\qquad(2.7)$$

which is an nth order difference equation in the number-
of-births variable x_0. Since, for all populations, there exists some age beyond which no individuals are known to

survive, equation (2.7) can always be invoked provided n is large enough.

Now that we have introduced both the mortality and natality components of the life table, it is useful to define the *net reproductive value*, R_0, and the *mean length of a generation*, T. Since each female, on entering the first age class is expected to produce $\sum_{i=1}^{n} b_i l_i$ females (assuming that n is large enough so that $s_n = 0$), the average individual net reproductive value is

$$R_0 = \sum_{i=1}^{n} b_i l_i. \tag{2.8}$$

For a general natality schedule, individuals reproduce repeatedly at several ages; that is, reproduction is distributed with age and the terms

$$\frac{b_i l_i}{R_0}$$

represent the relative reproduction frequencies with respect to age i. Thus, from probability theory, it follows that the average of this distribution, which is the average age of a mother when giving birth (more formally referred to as the mean length of a generation), is

$$T = \frac{\sum_{i=1}^{n} i b_i l_i}{R_0}. \tag{2.9}$$

Other useful quantities can be defined. Those relating to the notion of stable age distributions are introduced in the following section.

25

CHAPTER TWO

2.2 POPULATION DYNAMICS

2.2.1 *Leslie Matrix Model*

From equation (2.3), it follows that $x_1(t+1) = s_0 x_0(t)$. Applying equation (2.5) at time t rather than time $t+1$, it follows that $x_1(t+1)$ can be expressed as

$$x_1(t+1) = s_0 \sum_{i=1}^{n} b_i x_i(t). \tag{2.10}$$

It now follows from equations (2.3) and (2.10) that the n-dimensional state space vector $\mathbf{x}(t) = (x_1(t), x_2(t), \ldots, x_n(t))'$ (notationally this n-dimensional space will be written as R^n) satisfies the matrix equation

$$\mathbf{x}(t+1) = L\mathbf{x}(t), \tag{2.11}$$

where the so-called Leslie matrix L has the form

$$L = \begin{pmatrix} s_0 b_1 & s_0 b_2 & \cdots & s_0 b_{n-1} & s_0 b_n \\ s_1 & 0 & \cdots & 0 & 0 \\ 0 & s_2 & \cdots & 0 & 0 \\ \vdots & \vdots & \ddots & \vdots & \vdots \\ 0 & 0 & \cdots & 0 & 0 \\ 0 & 0 & \cdots & s_{n-1} & s_n \end{pmatrix}. \tag{2.12}$$

The above formulation differs from the original formulation of Leslie (1945) in excluding the number of births variable x_0 from equation (2.11) and allowing $s_n \neq 0$ (see Emlen, 1984, for a treatment that explicitly includes x_0). This model was also derived by Lewis (1942), while Lefkovitch (1965, 1967) and Usher (1966, 1976) pioneered its application to nonhuman populations and resource management.

In some of the ensuing analyses we assume that $s_n = 0$; but, for the purposes of generality, s_n is retained in (2.12) because it is nonzero in, for example, applications to certain fish populations where individuals beyond a particular

26

age are assumed to have the same mortality and fecundity rates.

The natality and mortality life table parameters can be used to calculate the structure of the stationary age distribution, intrinsic growth rate, and mean generation time for a population. By considering the dynamic equation (2.10), however, a more general nonstationary theory can be developed that allows one to deduce under what conditions such exotic dynamic behavior exists as a population cycle (for example, see Tuljapurkar, in press).

2.2.2   General Linear Model

The Leslie matrix equation (2.11) is a special case of the more general system of linear matrix equations

$$\mathbf{x}(t + 1) = A\mathbf{x}(t), \tag{2.13}$$

where A is an $n \times n$ matrix of real (as opposed to complex) elements a_{ij}, $i, j = 1, \ldots, n$. The general theory associated with equation (2.13) can be found in many texts (Luenberger, 1979, provides a particularly lucid treatment). For the sake of completeness we outline those results which are used to develop the theory presented below. The behavior of solutions to equation (2.13) is critically dependent on the eigenvalues of A, that is, the set of values λ that satisfy the algebraic equation

$$A\mathbf{x} = \lambda\mathbf{x}, \tag{2.14}$$

for some vector $\mathbf{x}$. This vector is termed the *right* eigenvector corresponding to the eigenvalue λ. A set of *left* eigenvectors is obtained by considering the algebraic equations

$$\mathbf{x}'A = \lambda\mathbf{x}'. \tag{2.15}$$

Note that both equations (2.14) and (2.15) remain true if each element of the eigenvectors is multiplied by the same constant. (An eigenvector determines a unique direction in R^n given by the proportional or normalized values of its elements, but the magnitude is arbitrary.) Also, both equations imply that the eigenvalues satisfy the *characteristic polynomial* in λ obtained by expanding the expression

$$\det[A - \lambda I] = 0, \tag{2.16}$$

where $\det[\cdot]$ is the determinant of the matrix enclosed within the brackets. Such polynomials are known to have n roots λ_i, $i = 1, \ldots, n$ (not all are necessarily distinct and some or all may be complex), which are often represented by the n-dimensional diagonal matrix

$$\Lambda = \begin{pmatrix} \lambda_1 & 0 & \cdots & 0 \\ 0 & \lambda_2 & \cdots & 0 \\ \vdots & \vdots & \ddots & \vdots \\ 0 & 0 & \cdots & \lambda_n \end{pmatrix}.$$

A matrix A is said to be diagonalizable if there exists an $n \times n$ nonsingular matrix T (with inverse denoted by T^{-1}) and diagonal matrix Λ such that

$$A = T\Lambda T^{-1}. \tag{2.17}$$

Starting from the initial condition $\mathbf{x}(0)$, it follows from (2.17) and by iterating relationship (2.16) that, in cases where A can be diagonalized, the solution to equation (2.13) is given by

$$\mathbf{x}(t + 1) = A^t \mathbf{x}(0) = T\Lambda T^{-1} T\Lambda T^{-1} \ldots T\Lambda T^{-1} \mathbf{x}(0);$$

that is (since by definition $T^{-1}T = I$, where I is the identity matrix),

$$\mathbf{x}(t + 1) = T\Lambda^t T^{-1} \mathbf{x}(0), \tag{2.18}$$

where

$$\Lambda^t = \begin{pmatrix} \lambda_1^t & 0 & \cdots & 0 \\ 0 & \lambda_2^t & \cdots & 0 \\ \vdots & \vdots & \ddots & \vdots \\ 0 & 0 & \cdots & \lambda_n^t \end{pmatrix}.$$

Thus it follows, for all $i = 1, \ldots, n$, that if the magnitudes (modulus for complex numbers, absolute value for real numbers) of λ_i, denoted by $|\lambda_i|$, are less than 1, then from equation (2.18) all elements of the solution $\mathbf{x}(t) \rightarrow 0$ as $t \rightarrow \infty$, and the point 0 is referred to as being *asymptotically stable*. Since this property does not depend on the initial value of $\mathbf{x}(0)$, the point 0 is more exactly referred to as being *globally asymptotically stable*. If $|\lambda_i| > 1$ for one or more values of $i = 1, \ldots, n$, however, then some or all of the elements of $\mathbf{x}$ grow without bound and the system is referred to as being *unstable*. In fact, if λ_Δ is the dominant eigenvalue (largest in magnitude), then the solution will ultimately grow in multiples of λ_Δ and align itself in the direction of the corresponding eigenvector $\mathbf{x}_\Delta$ (see Luenberger, 1979, Section 5.11, for more details).

An equilibrium solution is one that satisfies

$$\mathbf{x}(t) = \mathbf{x}, \quad \text{a constant for all values of } t. \tag{2.19}$$

It thus follows from eigenvalue equation (2.14) that eigenvectors corresponding to eigenvalues having the value 1 are equilibrium solutions. If this is not the case, then the only equilibrium solution is the trivial point $\mathbf{x} = 0$, and its stability properties are as discussed above.

From equations (2.13) and (2.14) it follows that all eigenvectors have the property that

$$\mathbf{x}(t + 1) = \lambda \mathbf{x}(t),$$

or

$$\mathbf{x}(t + 1) = \lambda^{t+1} \mathbf{x}(0),$$

provided $x(0)$ is the eigenvector corresponding to a particular value of λ. Thus for any initial condition $x(0)$ that corresponds to a *real* eigenvector, solutions grow ($\lambda > 1$), decrease ($0 < \lambda < 1$), or oscillate with decreasing ($-1 < \lambda < 0$) or increasing ($\lambda < -1$) amplitude. We know, as discussed above, for noneigenvector initial conditions, $x(0)$, that solutions ultimately align themselves in the direction of the dominant eigenvector x_Δ that corresponds to the dominant eigenvalue λ_Δ.

2.2.3 Positive Linear Models

As in the Leslie matrix model, physical or biological considerations often imply that all the elements of a particular matrix A in a population equation of the form (2.13) are nonnegative. The inequality $A \geq 0$ will be used to denote such matrices, where we will always assume that at least one element of A is positive. For any two vectors x and y of the same dimension, we will use the notation $x \geq y$ to mean that the ith elements of these vectors satisfy $x_i \geq y_i$ for all values of i. If the initial condition satisfies $x(0) \geq 0$, then it follows from equation (2.13) that whenever $A \geq 0$ the solution $x(t)$ will satisfy $x(t) \geq 0$ for all $t = 1, 2, \ldots$ (note, depending on the context, that 0 is used to denote the null matrix, null vector, or scalar value zero). Thus in models satisfying $A \geq 0$, the biological interpretation of "numbers of individuals," assigned to each element of $x(t)$, remains valid as the system modeled by equation (2.13) evolves through time.

We now state without proof, the *Perron-Frobenius theorem*, a result that is central to our understanding of the behavior of linear population models (for a proof see Luenberger 1979, Section 6.2).

Theorem 2.1 (Perron-Frobenius) *If $A \geq 0$ and all elements of A^t are strictly positive for some value of t (a positive integer),* *then there exists an eigenvalue λ_Δ of A (Perron root) that is positive and larger in magnitude than the remaining $n - 1$ eigenvalues.*

Since $A \geq 0$ and its dominant eigenvalue is positive, in order for equation (2.14) to hold, the corresponding dominant eigenvector x_Δ must have nonnegative elements.

The importance of the Perron-Frobenius theorem lies in the fact that for populations modeled by equation (2.13), it guarantees the existence of a *stable population structure*, x_Δ, and *growth rate*, λ_Δ, that determine the ultimate dynamical behavior of the population; that is, as $t \to \infty$, $x(t)$ aligns itself in the same direction as x_Δ and changes in magnitude at a rate that approaches λ_Δ.

2.2.4   Stage-structured Populations

Life tables are generally based on age structure, and the Leslie model given by equation (2.11) assumes that each individual moves up one age class in one time period. For most biological populations, the age of individuals is difficult to determine. Other quantities often correlated with age, such as size or weight, may be more convenient to measure and more pertinent to questions relating to resource exploitation. These qualities can be used to structure populations into classes that we will refer to as "stage classes." We will assume that the classes can be ordered (numbered) such that, for a small enough time step, each individual either remains in that class or proceeds into the next class. If the growth of individuals in a population is monotonic (as is the case for many populations if seasonal fluctuations and very old individuals are ignored), then weight or size classes satisfy our assumptions provided that the ranges of values in each class are suitably large (large enough to prevent an individual from skipping over a stage class in one time period).

Let $0 < p_i \leq 1$, $i = 1, \ldots, n - 1$, denote the proportion of individuals in stage class i at time t that move into the next stage class at time $t + 1$ (by definition $p_n = 0$). Thus $1 - p_i$ is the proportion of individuals in stage class i that remain behind. As before, we can define survival rates s_i, $i = 1, \ldots, n$. However, these no longer have the age interpretation associated with equation (2.1) but the more general interpretation of the fraction of individuals in stage class i that survive the time interval $(t, t + 1)$. Equation (2.3) therefore generalizes to

$$x_{i+1}(t + 1) = (1 - p_{i+1})s_{i+1}x_{i+1}(t) + p_i s_i x_i(t),$$
$$i = 1, \ldots, n - 1. \qquad (2.20)$$

Assume further that the birth process is still defined by equation (2.5), but that the coefficients b_i now have the interpretation of newborn females per female in stage class i. Then the number of individuals that enter stage class 1 at time $t + 1$ is the number of individuals born at time t that survive to time $t + 1$ (that is, $s_0 x_0(t)$) plus the number of individuals already in the stage class at time t that survive to time $t + 1$ but do not make the transition to stage class 2 at time $t + 1$ (that is, $(1 - p_1)s_1 x_1$). Hence defining $p_0 = 1$, equations (2.21) also hold for $i = 0$; that is,

$$x_1(t + 1) = (1 - p_1)s_1 x_1(t) + s_0 x_0(t), \qquad (2.21)$$

where $x_0(t) = \sum_{i=1}^{n} b_i x_i(t)$ (as given by equation (2.5)).

Note that a model constructed using parameters p_i, as defined above, implicitly assumes that all individuals in a stage class are equivalent, regardless of the number of time periods that any individual has previously spent in that stage class. This is an obvious weakness of the model, but it is a weakness that is inherent in any discrete-size structure model (continuous models suffer from other simplifying assumptions about the rates of processes as a function of size or age). The utility of this stage-equivalency assumption depends, of course, on the particular application. For insect populations, as discussed in Chapter 6, models have been formulated that keep both age and stage classes separate (for example, see Chi and Liu, 1985). These models, however, require many more parameters than the stage-class model presented here, and also require measuring the age of individuals. When it comes to the problem of estimating model parameters from biological data, it is known that the model with the most parameters does not always more accurately predict population growth (for example, see Hochberg *et al.*, 1986). Further, it is known that the more simple models may often lead to insights not gained from very detailed models.

The stage-class model (2.20) was first introduced into resource management by Usher (1966). It includes the Leslie matrix model as a special case (set $p_i = 1$, $i = 1, \ldots, n - 1$, $P_n = 0$) and appears to be particularly well suited to dealing with forest-stand management problems.

Clearly equations (2.20) can be written in the matrix form of equation (2.13), where the transformation matrix is

$$A = \begin{pmatrix} (1 - p_1)s_1 + s_0 b_1 & \cdots & s_0 b_{n-1} & s_0 b_n \\ p_1 s_1 & \cdots & 0 & 0 \\ \vdots & \ddots & \vdots & \vdots \\ 0 & \cdots & (1 - p_{n-1})s_{n-1} & 0 \\ 0 & \cdots & p_{n-1}s_{n-1} & s_n \end{pmatrix}. \tag{2.22}$$

Since this matrix A is nonnegative, the Perron-Frobenius theorem applies. The Perron root and all other eigenvalues satisfy equation (2.16) which, using the standard theory of expanding determinants (Laplace's expansion; for example, see Luenberger, 1979, Section 3.2), leads to the result stated in the following lemma.

Lemma 2.2 *The characteristic equation for the eigenvalues as-*

sociated with the matrix A defined in (2.22) is the n*th order polynomial,*

$$\prod_{j=1}^{n} \left(\lambda - (1-p_j)s_j \right) - \sum_{i=1}^{n} s_0 b_i \prod_{j=1}^{i-1} p_j s_j \prod_{j=i+1}^{n} \left(\lambda - (1-p_j)s_j \right) = 0.$$

(Note that we use the convention $\prod_{j=1}^{0}(\cdot) = 1$ and also, when $i = n$, $\prod_{j=n+1}^{n}(\cdot) = 1$.) Provided $(\lambda - (1 - p_j)s_j) \neq 0$ for some $j = 1, \ldots, n$, this characteristic equation can be rewritten as

$$\sum_{i=1}^{n} \frac{s_0 b_i}{p_i s_i} \prod_{j=1}^{i} \frac{p_j s_j}{\lambda - \left(1 - p_j\right) s_j} = 1. \tag{2.23}$$

This equation, as will be demonstrated later, is particularly useful for the purposes of analyzing the properties of the eigenvalues associated with the special case $A = L$, where L is the Leslie matrix (2.12).

We know from the Perron-Frobenius theorem that a positive solution λ_Δ exists and dominates the magnitude of all other solutions to equation (2.23). In the case of the Leslie matrix model ($p_i = 1$, $i = 1, \ldots, n - 1$), a stronger result can be derived, as discussed below.

We now present equations (2.20) and (2.21) in a notational form that facilitates the development and analysis of the nonlinear extensions used to model fisheries and forest-stand management problems. At first, the approach might seem a little strange to those familiar with Leslie matrix theory, but its value will become apparent when we discuss the nonlinear equilibrium analysis in Chapter 3. Define the $n \times n$ survival matrix S as

$$S = \begin{pmatrix} s_1 & 0 & \cdots & 0 \\ 0 & s_2 & \cdots & 0 \\ \vdots & \vdots & \ddots & \vdots \\ 0 & 0 & \cdots & s_n \end{pmatrix}, \tag{2.24}$$

34

and the $n \times n$ transition matrix P (recalling that $p_n = 0$ by definition) as

$$P = \begin{pmatrix} 1 - p_1 & 0 & \cdots & 0 & 0 \\ p_1 & 1 - p_2 & \cdots & 0 & 0 \\ \vdots & \vdots & \ddots & \vdots & \vdots \\ 0 & 0 & \cdots & 1 - p_{n-1} & 0 \\ 0 & 0 & \cdots & p_{n-1} & 1 \end{pmatrix}. \quad (2.25)$$

If we define an input vector $(s_0 x_0(t), 0, \ldots, 0)'$ and a stage-progression matrix $G = PS$, then from the definitions (2.24) and (2.25) it follows (recalling $p_n = 0$) that

$$G = \begin{pmatrix} (1 - p_1)s_1 & \cdots & 0 & 0 \\ p_1 s_1 & \cdots & 0 & 0 \\ \vdots & \ddots & \vdots & \vdots \\ 0 & \cdots & (1 - p_{n-1})s_{n-1} & 0 \\ 0 & \cdots & p_{n-1}s_{n-1} & s_n \end{pmatrix}, \quad (2.26)$$

and equations (2.20) and (2.21) can be written as

$$\mathbf{x}(t + 1) = G\mathbf{x}(t) + \begin{pmatrix} s_0 x_0(t) \\ 0 \\ \vdots \\ 0 \end{pmatrix}. \quad (2.27)$$

Note that although the vector $(s_0 x_0(t), 0, \ldots, 0)'$ can be eliminated from equation (2.27) by incorporating the expression

$$s_0 x_0(t) = s_0 \sum_{i=1}^{n} b_i x_i(t)$$

(see equation (2.5)) into an expanded matrix G (which is just the matrix A defined in (2.22)), there are several advantages to explicitly identifying $x_0(t)$ in equation (2.27). As previously mentioned, these advantages will become apparent in the development of the nonlinear theory in the next chapter. Also G, as it stands, represents a *progression/mortality* process in the population as individuals move through the stage classes, while $(s_0 x_0(t), 0, \ldots, 0)'$ can be thought of as a *driving vector* that can be generalized to include migration (for example, if the ith element of the driving vector is nonzero, then there is a net immigration [positive element] or emigration [negative element] of individuals into or out of the ith stage class).

To facilitate analysis of equation (2.27) in the nonlinear setting developed in the next chapter, it is useful to characterize the inverse of lower diagonal matrices like G, that is, matrices whose only nonzero elements are on the diagonal and lower subdiagonal. Using forward substitution to solve the system of linear equations $\mathbf{y} = G\mathbf{x}$, it is easily shown that the elements of the inverse matrix G^{-1} have the form given in the following lemma.

Lemma 2.3 *If a matrix G with elements g_{ij}, $i, j = 1, \ldots, n$ satisfies $g_{ii} \neq 0$ for $i = 1, \ldots, n$ and $g_{ij} = 0$ for all i, and for $j \neq i, i-1$, then the ijth element of the inverse matrix G^{-1} satisfies*

$$\left(G^{-1}\right)_{ij} = \begin{cases} 0, & \text{if } i < j; \\ \dfrac{(-1)^{i-j}}{g_{i+1\,i}} \displaystyle\prod_{r=j}^{i} \left(\dfrac{g_{r+1\,r}}{g_{rr}}\right), & \text{if } i \geq j \end{cases} \qquad (2.28)$$

Note that when $r = i$ in the above product the term $g_{i+1\,i}$ cancels with the term outside the product so that the undefined quantity $g_{n+1\,n}$ does not really occur when $i = n$. Also when $i = j$, the right-hand side of equation (2.28) reduces to $1/g_{ii}$. Note that an equivalent result to (2.28) also holds for upper diagonal matrices.

Suppose $\mathbf{x}_\Delta$ is the right eigenvector associated with the Perron root λ_Δ of the matrix A defined by (2.22). Then it follows from equation (2.27) that (see equation (2.14))

$$\lambda_\Delta \mathbf{x}_\Delta = G\mathbf{x}_\Delta + \begin{pmatrix} s_0 x_0 \\ 0 \\ \vdots \\ 0 \end{pmatrix},$$

36

where x_0 is determined from equation (2.5) using $\mathbf{x}_\Delta$ instead of $\mathbf{x}(t+1)$. This equation can be rearranged to obtain

$$(\lambda_\Delta I - G)\mathbf{x}_\Delta = \begin{pmatrix} s_0 x_0 \\ 0 \\ \vdots \\ 0 \end{pmatrix},$$

which has the solution

$$\mathbf{x}_\Delta = (\lambda_\Delta I - G)^{-1} \begin{pmatrix} s_0 x_0 \\ 0 \\ \vdots \\ 0 \end{pmatrix}, \qquad (2.29)$$

provided the matrix $(\lambda_\Delta I - G)$ is invertible. Thus all the elements of $\mathbf{x}_\Delta$ can be expressed in terms of the variable x_0. Since this matrix is also a lower diagonal matrix, Lemma 2.3 can be applied to obtain the following result.

Result 2.4 *The ijth element of the matrix $(\lambda_\Delta I - G)^{-1}$, where G is defined by (2.26), has the form*

$$\left((\lambda_\Delta I - G)^{-1} \right)_{ij} = \begin{cases} 0, & \text{if } i < j; \\ \frac{1}{p_i s_i} \prod_{r=j}^{i} \frac{p_r s_r}{\lambda_\Delta - (1-p_r)s_r}, & \text{if } i \geq j. \end{cases}$$

This result can now be applied to equation (2.29) to explicitly obtain the vector $\mathbf{x}_\Delta$ in terms of x_0. As discussed after Theorem 2.1, one consequence of the Perron-Frobenius theorem is that $\mathbf{x}_\Delta$ is nonnegative. Hence it is the *stable stage distribution* since solutions starting at all initial conditions (except those that correspond to other eigenvalues) approach the same stage distribution as $\mathbf{x}_\Delta$. If we normalize $\mathbf{x}_\Delta$ so that the sum of its elements add to unity, then this normalized vector represents the proportion of individuals in each stage class of the stable stage distribution. Thus if we define c_i, $i = 1, \ldots, n$, to be this proportion, then it follows, applying Result 2.4 to equation (2.29), that

$$c_i = \left(\frac{1}{p_i s_i} \prod_{j=1}^{i} \frac{p_j s_j}{\lambda_\Delta - (1 - p_j) s_j} \right) \bigg/ \kappa, \qquad (2.30)$$

where

$$\kappa = \sum_{i=1}^{n} \frac{1}{p_i s_i} \prod_{j=1}^{i} \frac{p_j s_j}{\lambda_\Delta - (1 - p_j) s_j}. \qquad (2.31)$$

The latter expression ensures that $\sum_{i=1}^{n} c_i = 1$.

We also know that there is a left eigenvector associated with the Perron root of the stage-structured model (see equation (2.15)). If this eigenvector is denoted by $\mathbf{x}_{\Delta'}$, then it follows from equation (2.15) that

$$\lambda_\Delta \mathbf{x}_{\Delta'} = A' \mathbf{x}_{\Delta'},$$

where A' is the transpose of the matrix defined in (2.22). Thus instead of satisfying an equation of the form (2.27), $\mathbf{x}_{\Delta'}$ satisfies the equation

$$\lambda_\Delta \mathbf{x}_{\Delta'} = G' \mathbf{x}_{\Delta'} + s_0 x_1 \begin{pmatrix} b_1 \\ \vdots \\ b_n \end{pmatrix}.$$

As in deriving equation (2.29) it now follows that $\mathbf{x}_{\Delta'}$ satisfies

$$\mathbf{x}_{\Delta'} = s_0 x_1 ((\lambda_\Delta I - G)^{-1})' \begin{pmatrix} b_1 \\ \vdots \\ b_n \end{pmatrix}.$$

Applying Result 2.4 to obtain explicit expressions for the elements of the left eigenvector $\mathbf{x}_{\Delta'}$, it is easily shown that its jth element is proportional to the quantity

$$v_j = \sum_{i=j}^{n} \frac{s_0 b_i}{p_i s_i} \prod_{r=j}^{i} \frac{p_r s_r}{\lambda_\Delta - (1 - p_r) s_r}, \quad j = 1, \dots, n. \quad (2.32)$$

Since for any set of terms $\{\tau_r | r = 1, \ldots, n\}$ and indices $1 < j \le i \le n$, the product $\tau_j \times \cdots \times \tau_i$ can be expressed as

$$\prod_{r=j}^{i} \tau_r = \prod_{r=1}^{i} \tau_r \Bigg/ \prod_{r=1}^{j-1} \tau_r,$$

it follows from equation (2.30) that

$$v_1 = \kappa s_0 \sum_{i=1}^{n} b_i c_i$$

$$v_j = \frac{\kappa s_0}{p_{j-1} s_{j-1} c_{j-1}} \sum_{i=j}^{n} b_i c_i, \quad j = 2, \ldots, n. \quad (2.33)$$

Since the elements of the vector $\mathbf{v} = (v_1, \ldots, v_n)'$ are proportional to $\mathbf{x}_{\Delta'}$, $\mathbf{v}$ is a left eigenvector corresponding to the Perron root λ_Δ. The elements of this eigenvector, like the stable stage-structure right eigenvector $\mathbf{c} = (c_1, \ldots, c_n)'$ (whose elements are proportional to $\mathbf{x}_\Delta$), also have a biological interpretation. First note from (2.23) and (2.32) that $v_1 = 1$. Therefore it follows from (2.33) that $\kappa s_0 = 1 / \sum_{i=j}^{n} b_i c_i$. Thus κs_0 is the reciprocal of the average number of births per individual in the stable stage-distributed population (note that the elements c_i are a stage-frequency distribution). Furthermore, since the quantity $\kappa s_0 \sum_{i=j}^{n} b_i c_i$ is the proportion of progeny that a newborn individual expects to contribute during the period it is a member of stage classes j to n, and $p_{j-1} s_{j-1} c_{j-1}$ is the proportion of individuals in stage $j-1$ that make the transition to the next stage class each time period, it follows that v_j, defined by expression (2.33), is the relative reproductive value of individuals as they enter stage class j from stage class $j - 1$. This definition represents the generalization of Fisher's *reproductive value* for age-structured populations (R. A. Fisher, 1930) to stage-structured populations (cf. equation (2.8)). We can also

39

generalize Fisher's concept of *total reproductive value* by defining the same scalar quantity $V(t) = \mathbf{v}'\mathbf{x}(t)$, where $\mathbf{x}(t)$ is now the stage state rather than the age state of the population at time t. As for age-structured populations, since $\mathbf{v}$ is a left eigenvalue, it follows that

$$V(t+1) = \mathbf{v}'\mathbf{x}(t+1) = \lambda_\Delta \mathbf{v}'\mathbf{x}(t) = \lambda_\Delta V(t); \qquad (2.34)$$

that is, the total reproductive value of a stage-structured population grows at a rate determined by its Perron root (or decreases if the root is less than 1).

2.2.5 Age-structured Populations

We return to the Leslie matrix model which, as has been mentioned already, is a special case of the stage-structured model with $p_i = 1$, $i = 1, \ldots, n-1$. Thus it follows from the results of the previous section that L defined in (2.12) (with $s_n = 0$) has a Perron root; but a stronger result can be obtained, as derived below, if we assume that at least two consecutive age classes are fecund (that is, $b_i > 0$ and $b_{i+1} > 0$ for some $i = 1, \ldots, n-1$). More generally (see Sykes, 1969; Demetrius, 1971, 1972; Emlen, 1984), the result stated below remains true if the greatest common divisor of the indices j for which $b_j > 0$ is 1. Furthermore, if this greatest common divisor is greater than 1, then population cycles are known to exist. Thus consider equation (2.23) for the case $A = L$. Using relationship (2.2), it follows that equation (2.23) reduces to

$$\sum_{i=1}^{n} \frac{b_i l_i}{\lambda^i} = 1, \qquad (2.35)$$

provided $\lambda \neq 0$.

An analysis of the properties of equation (2.35) leads to the following result.

Result 2.5 *Consider a population modeled by the Leslie equation (2.11) (assuming $s_n = 0$). If the greatest common divisor*

40

*of the age index i is 1 for those age classes for which $b_i \neq 0$,
then an age distribution exists that is approached asymptotically
(even though the population itself may be approaching 0 asymptotically) for all $\mathbf{x}_0 \geq 0$. Furthermore, the population will grow
without bound or will asymptotically approach 0 depending on
whether the value of $\sum_{i=1}^{n} b_i l_i$ is respectively greater or less than
1.*

Note that this result applies if any two consecutive age
classes of females has nonzero fecundity, but not, for
example, to a population for which only b_4 and b_6 are
nonzero (the lowest common divisor of 4 and 6 is 2). Here
we will prove the result for the consecutive age-class case.
The more general result follows by appropriately modifying this proof (also see Demetrius, 1971).

Proof. Since $s_i > 0$, $i = 1, \ldots, n - 1$ (otherwise the number of age classes would be less than n), it follows from
expression (2.2) that $l_i > 0$, $i = 1, \ldots, n$. Thus the coefficients of the polynomial in λ on the right-hand side of
equation (2.35) are nonnegative. Hence this polynomial
is a strict monotonically increasing function for $\lambda > 0$,
thereby taking on the value 1 for a unique positive value
of λ. This implies that equation (2.35) has exactly one positive real root, which we denote by λ_Δ. Also the nonnegativity of all the coefficients implies that $\lambda_\Delta > (<)1$ whenever $\sum_{i=1}^{n} b_i l_i > (<)1$. Hence from our earlier discussion
on the stability of matrix equation (2.13), the stability result follows once we have proven that the eigenvalue λ_Δ
has a larger modulus than all other eigenvalues.

Consider the eigenvalues λ represented in complex po-
lar coordinates by $(1/\lambda) = re^{i\theta}$, where $r > 0$, $0 \leq \theta \leq 2\pi$
and $\mathbf{i} = \sqrt{-1}$. Since λ_Δ is real ($\theta_\Delta = 0$), equation (2.35)
can be written as

$$\sum_{i=1}^{n} b_i l_i (r_\Delta)^i = 1.$$

For the remaining eigenvalues, which are either complex ($\theta_\Delta \neq 0$ *or* π) or negative ($\theta_\Delta = \pi$), the real part of this equation can be written as

$$\sum_{i=1}^{n} b_i l_i r^i \cos i\theta = 1.$$

Since $|\cos i\theta| \leq 1$, it follows from the monotonicity of the polynomial in λ that $r \geq r_\Delta$, with equality only holding if and only if all cosine terms with nonzero coefficients are unity. For strictly complex roots (that is, $\theta_\Delta \neq 0$ or π), it is impossible for both consecutive terms $\cos i\theta$ and $\cos(i+1)\theta$ to be unity. But, by assumption, at least two (consecutive) parameters b_i and, hence, coefficients $b_i l_i$, are positive. It thus follows that $r > r_\Delta$ or $(1/r) < (1/r_\Delta)$ for all complex roots λ, and λ_Δ is larger in magnitude than all other eigenvalues. Further, from eigenvalue equation (2.14), if $x \geq 0$ and $A \geq 0$, then λ is nonnegative. Hence, no nonnegative eigenvectors exist besides x_Δ and all solutions beginning with initial conditions satisfying $x_0 \geq 0$ must approach the eigenvector corresponding to the dominant eigenvalue. $\square$

The structure of the stable age distribution follows by setting $p_i = 1$, $i = 1, \ldots, n-1$ in equation (2.30) and (2.31); that is, we can state the following result.

Result 2.6 *The proportion of individuals, c_i, in the stable age distribution associated with the Leslie matrix model defined by equations (2.11) and (2.12) (with $s_n = 0$) is given by*

$$c_i = \frac{\lambda_\Delta^{-i} l_i}{\sum_{j=1}^{n} \lambda_\Delta^{-j} l_j}, \quad i = 1, \ldots, n, \qquad (2.36)$$

where λ_Δ is the Perron root associated with the eigenvalue equation (2.35).

42

Note that if $s_n \neq 0$, the last equation in (2.36) is

$$c_n = \left(\frac{\lambda_\Delta^{-i} l_i}{\sum_{j=1}^{n} \lambda_\Delta^{-j} l_j} \right) \left(\frac{\lambda_\Delta}{\lambda_\Delta - s_n} \right).$$

Recalling that $l_0 = 1$ and $l_i = \prod_{j=0}^{i-1} s_j$, $i = 1, \ldots, n$, (equation (2.2)), it follows that v_j, defined by identities (2.32), has the following form when $p_i = 1$, $i = 1, \ldots, n$:

$$v_j = \sum_{i=j}^{n} \frac{s_0 b_i l_i}{l_j \lambda_\Delta^{(i+1)-j}}. \tag{2.37}$$

Also recall that (l_i/l_j) is the probability of surviving from age j to age i (cf. equation (2.1)) and note from equation (2.34) that λ_Δ is the rate at which the total reproductive value of the population increases in each time period (or for that matter, it is the rate at which the size of the population grows when it is in its stable stage configuration). Hence $1/\lambda^r$ represents the relative change, at any time t, in the reproductive value of a single individual relative to the total population as the total population changes in size over the next r time periods. Thus, from equation (2.37), v_j represents a relative discounted value (relative to newborn individuals since $v_1 = 1$; cf. equation (2.35)) on the future schedule of offspring to a j-aged individual (also see Emlen, 1984).

2.3   HARVESTING THEORY

2.3.1   Formulation

Both the Leslie matrix and the more general stage-structured model defined by equation (2.27) have been used to address questions relating to optimal harvesting of populations. Essentially, there is no difference in the analysis between harvesting age- and stage-structured popula-

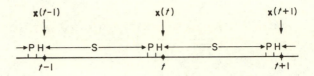

FIGURE 2.2. Convention for sequencing the survival (S), harvesting (H), and transition processes (P). Note that, under this convention, the variable $\mathbf{x}(t)$ has the interpretation of a *residual* state vector.

tions so that, for generality, we develop most of the theory in the stage-structured setting.

Since our model is discrete, we need to specify exactly when harvesting occurs with respect to the processes of aging (application of S defined in expression (2.24)), transition (application of P defined in expression (2.25)), and reproduction (addition of the term $s_0 x_0$ in equation (2.20) when $i = 1$). The convention depicted in Figure 2.2 is the one we adopt. It corresponds to the case of harvesting *after* rather than *before* reproduction. Also it is the more convenient and, as proved by Doubleday (1975), it is the more efficient of the two options (produces a higher yield because we allow individuals to reproduce just before harvesting). There is no loss of generality using this convention, since other conventions can be used to obtain the same set of qualitative results, although the particulars will differ with respect to the parameters that appear in various expressions (relating, say, to yield or stability calculations). It is important when estimating system parameters in applied problems, however, to ensure that the interpretation of data is consistent with the convention used to set up the dynamic equations. As indicated in Figure 2.2, at each discrete point in time the variable $\mathbf{x}(t)$ has the interpretation of a *residual* state vector of individuals that age over the right open interval $[t, t + 1)$ and then, just prior to time $t + 1$, undergo stage transition and harvesting in that order, where the ith element $u_i(t)$ of the vector $\mathbf{u}(t)$ denotes the number of individuals removed from the ith stage class to obtain the residual ith stage class $x_i(t+1)$ (one cannot harvest more animals than there are).

Under these assumptions, it follows that the stage-structured model (2.13), with the system matrix A as defined by (2.22), can be written as

$$\mathbf{x}(t+1) = A\mathbf{x}(t) - \mathbf{u}(t). \tag{2.38}$$

For consistency, we require that

$$\mathbf{0} \le \mathbf{u}(t) \le A\mathbf{x}(t) \quad i = 1, \dots, n \tag{2.39}$$

to ensure that $\mathbf{x}(t+1) \ge \mathbf{0}$.

The first step in any harvesting analysis is to formulate an appropriate question relating to the exploitation of a population modeled by equation (2.38). This is more difficult than it seems at first appearances, because a number of issues arise that cannot be resolved immediately. For example, suppose we are concerned simply with the question of choosing a set of harvesting strategies to maximize the value of the yield harvested over some time period $[0, T]$ (harvesting in years 1 to T), that is, to maximize the performance criterion

$$J_T(\mathbf{x}_0) = \sum_{t=0}^{T-1} \sum_{i=1}^{n} w_i(t)u_i(t) = \sum_{t=0}^{T-1} \mathbf{w}'(t)\mathbf{u}(t), \tag{2.40}$$

where the elements $w_i(t)$ of the vector $\mathbf{w}(t)$ represent the value, either in dollars (J_T is the accumulated dollar value) or biomass units (J_T is the accumulated biomass yield), of an individual in the ith stage class harvested at time t.

It is clear, as indicated by its argument and subscript, that the value of J, defined by expression (2.40), depends on the length of the planning period T and on the initial state of the population $\mathbf{x}_0$. This state $\mathbf{x}_0$ is given (or estimated by sampling the population at time $t = 0$), but the parameter T is free. Thus we need to decide what value to assign to T, that is, to decide on the length of our planning horizon. In addition, we need to decide whether some type of constraint should be placed on the number of individuals in total or in specific stage classes at the end of the planning horizon. Choosing not to constrain $\mathbf{x}(T)$ is equivalent to placing zero value on the state of the population at time T. This invariably leads to the population being overexploited, or possibly even decimated over the planned harvesting period. The solution to a maximum yield policy thus depends on how the problem is formulated, so that the "optimal solution" to an inadequately formulated problem may actually turn out to be a very poor solution with respect to the long-term management of a population.

2.3.2 Equilibrium Analysis

One of the questions that we will focus on in the next section is the proper formulation of finite planning horizon problems. A first step towards understanding these problems is to consider the sustainable yield or equilibrium management problem. Here we avoid the dynamic aspects of the problem by examining harvest policies that maintain the population in the same state from one time interval to the next, that is, in equilibrium. This also obviates the problem of selecting a value T for the planning horizon and endpoint condition $\mathbf{x}(T)$ or appropriate endpoint constraints. In this context, it is useful to characterize $\mathbf{u}$ in terms of the proportion of individuals rather than the absolute number removed from each stage class at the end of the transition interval. Thus we define

$$0 \leq h_i \leq 1, \ i = 1,\ldots,n \qquad (2.41)$$

to be the proportion of individuals removed from the ith stage class after the application of transformation A, and

46

define H to be a diagonal matrix with ith diagonal element h_i. Thus it follows that

$$\mathbf{u} = HA\mathbf{x}. \tag{2.42}$$

Note, as required from (2.42), that inequalities (2.39) are always satisfied when inequalities (2.41) hold, as they necessarily do by definition of h_i as a proportion.

Reorganizing equation (2.38) subject to the equilibrium condition (2.19) and definition (2.42), the *maximum sustainable yield* (MSY) problem of maximizing the annual equilibrium yield $\mathbf{w'u}$ (cf. expression (2.40)) can be formally posed as follows.

Problem 1 (Maximum Sustainable Yield)

$$\max_{\mathbf{x} \geq \mathbf{0}} J(\mathbf{x}) = \mathbf{w}'HA\mathbf{x}, \tag{2.43}$$

subject to the constraint equation

$$(I - H)A\mathbf{x} = \mathbf{x}. \tag{2.44}$$

This problem is not completely specified. Below we show that its solution depends *a priori* on specifying the number of births x_0, although the qualitative form or structure of the solution is independent of x_0.

Equation (2.44), the equilibrium constraint, is satisfied for nonzero $\mathbf{x} \geq \mathbf{0}$ if and only if the matrix $(I - H)A$ has an eigenvalue $\lambda = 1$. Since the elements of A are fixed, equation (2.44) places an additional constraint on the elements of H, which are already constrained by inequalities (2.41). Note that equation (2.44) has a solution for some H satisfying inequalities (2.41) only if the Perron root of A satisfies $\lambda_\Delta \geq 1$ (see Doubleday, 1975). The elements of H are thus chosen to reduce the Perron root of the modified matrix $(I - H)A$ to unity, whence the vector $\mathbf{x}$

satisfying equation (2.44) is the stable stage distribution of the harvested population.

If we recall from equation (2.5) that

$$x_0 = \sum_{i=1}^{n} b_i x_i,$$

then the constraint equation (2.44) can be eliminated and the MSY problem expressed in terms of x_0 as follows. Divide A (defined by matrix (2.22)) into the matrix G and the vector $(s_0 x_0, 0, \ldots, 0)$, as was done in equation (2.27). Equation (2.44) then reduces to (cf. equation (2.29))

$$\mathbf{x} = (I - (I - H)G)^{-1} \begin{pmatrix} s_0 x_0 \\ 0 \\ \vdots \\ 0 \end{pmatrix}, \qquad (2.45)$$

and since $I - H$ is a diagonal matrix with ith diagonal element $1 - h_i$, it follows in a manner analogous to the derivation of Result 2.4 that

$$\left((I - (I - H)G)^{-1} \right)_{ij} =$$
$$\begin{cases} 0, & \text{if } i < j; \\ \dfrac{1}{p_i s_i (1 - h_{i+1})} \displaystyle\prod_{r=j}^{i} \dfrac{p_r s_r (1 - h_{r+1})}{1 - (1 - p_r) s_r (1 - h_r)}, & \text{if } i \geq j \end{cases} \cdot$$
$$(2.46)$$

Using expression (2.46) to explicitly write out the elements of $\mathbf{x}$ in equation (2.45) we finally obtain

$$x_i = \frac{s_0 x_0 (1 - h_1)}{p_i s_i (1 - h_{i+1})} \prod_{j=1}^{i} \frac{p_j s_j (1 - h_{j+1})}{1 - (1 - p_j) s_j (1 - h_j)},$$
$$i = 1, \ldots, n, \qquad (2.47)$$

Note that for $i = n$, since $p_n = 0$ and the terms in the undefined quantity $1 - h_{n+1}$ cancel, we obtain

$$x_n = \frac{s_0 x_0 (1 - h_1)}{1 - s_n(1 - h_n)} \prod_{j=1}^{n-1} \frac{p_j s_j (1 - h_{j+1})}{1 - (1 - p_j) s_j (1 - h_j)}.$$

It can also be shown for $i = 1$ (as before, we define $\prod_{j=1}^{0}(\cdot) = 1$) that expression (2.47) simplifies to

$$x_1 = \frac{s_0 x_0 w_1 h_1}{1 - (1 - p_1) s_1 (1 - h_1)}.$$

Expression $J(\mathbf{x})$ in (2.43) now becomes

$$J = s_0 x_0 (1 - h_1) \sum_{i=1}^{n} w_i h_i \left(\frac{1}{1 - h_i} + \frac{(1 - p_i) s_i}{1 - (1 - p_i) s_i (1 - h_i)} \right)$$

$$\times \prod_{j=1}^{i-1} \frac{p_j s_j (1 - h_{j+1})}{1 - (1 - p_j) s_j (1 - h_j)}, \quad i = 1, \ldots, n. \tag{2.48}$$

Note that the value of J depends linearly on x_0, which is consistent with the fact that constraint equation (2.44) determines the direction of the stable stage distribution associated with this equation, but not its absolute value. In fact, one of the major difficulties with a linear formulation of a harvesting problem is that *ad hoc* criteria must be used to select a particular birth level x_0 before the optimal harvesting regime can be implemented. On the other hand, the variables h_i (satisfying inequalities (2.41)) that maximize J, as expressed in (2.48), are independent of x_0 (since x_0 is just a linear multiplying factor in expression (2.48)); and thus the qualitative structure of the MSY policy is independent of x_0 or, equivalently, of population size.

In the simpler age-structured setting (also using the convention of harvesting just after rather than before the beginning of each time period), Beddington and Taylor

(1973) examined this problem and concluded that if an optimal harvesting policy can be found, then one exists that involves harvesting two age classes at most; typically one age class is completely harvested ($h_i = 1$ for some i which obviously implies that no individuals survive beyond this age), and a younger age class is partially harvested. A similar result is also proven by Rorres and Fair (1975). We will derive the same result in the next section for stage-structured models in a nonlinear setting; but, from our proof, it will become obvious that this bimodal harvesting result associated with the MSY problem stated here relates to the fact that only one of the inequality constraints associated with $h_i \leq 1$, $i = 1, \ldots, n$ can be operative (see remark above), and that constraint equation (2.44) can always be used to express the elements of $\mathbf{x}$ in terms of the scalar variable x_0 (number of births).

2.3.3 Linear Programming

A problem, such as Problem 1 (MSY), that involves maximizing a linear function of several variables subject to a set of linear constraints is referred to as a linear programming (LP) problem. For the sake of completeness, we will present a general formulation of the LP problem and state without proving the Fundamental Existence Theorem of Linear Programming. This theorem is used in the next section to derive some results relating to the harvesting of density dependent stage-structured populations. Although we deal with positive matrices A, the statement below holds for any constant $m \times n$ matrix A and constant m-dimensional and n-dimensional vectors $\mathbf{b}$ and $\mathbf{w}$. Hence n is the number of variables and m is the number of constraints.

Problem 2 (Linear Programming) *Maximize*

$$J = \mathbf{w}'\mathbf{u}$$

subject to

$$Au \leq b$$

and

$$u \geq 0.$$

We now state the following theorem using nontechnical language (a technical version requires us to introduce the concepts of feasible, basic, and degenerate solutions). Note that this theorem is useful only when $m < n$.

Theorem 2.7 (Fundamental Existence Theorem) *If the LP problem has a solution, then it has a solution u^* in which at most m of its n elements are nonzero.*

Proofs of this theorem can be found in any text that provides a detailed treatment of the theory of linear programming.

2.3.4   Dynamic Analysis

Let us return to the problem of maximizing $J_T(x_0)$, defined by expression (2.40), subject to equation (2.38) holding for $t = 0, \ldots, T - 1$, and a given initial population state $x(0) = x_0$. Iterating equation (2.38) for $t = 0, \ldots, T - 1$, it follows that

$$x(T) = A^T x(0) - A^{T-1} u(0) - A^{T-2} u(t) - \cdots - u(T - 1).$$

Thus the dynamic problem reduces to maximizing $J_T(x_0)$, defined by expression (2.40), subject to

$$\sum_{t=0}^{T-1} A^t u(T - 1 - t) = A^T x(0) - x(T). \qquad (2.49)$$

Since this optimization problem does not attach any value to the state of the population at time T, it is clear (assuming $w'(T - 1) \geq 0$) that $J_T(x_0)$ is maximized by harvesting all individuals remaining at $t = T - 1$; that is, from

51

constraint (2.39), the optimal value $\mathbf{u}^\star(T-1)$ will always be determined by

$$\mathbf{u}^\star(T-1) = A\mathbf{x}(T-1),$$

with consequent optimal final state (see equation (2.38)) $\mathbf{x}^\star(T) = 0$.

To avoid annihilating the population at time T, we can constrain $\mathbf{x}(T) = \mathbf{x}_T$, where $\mathbf{x}_T$ is some minimal level to which we are willing to exploit the population. This level is *ad hoc*, as is the value of T, since both are unrelated in any way to the dynamics of the system determined by equation (2.38), or the economics of the problem expressed in (2.40). Once we have decided on values for T and $\mathbf{x}(T)$ then, given $\mathbf{x}(0)$, the right-hand side of equation (2.49) is determined and this equation represents an n-dimensional constraint on the concatenated nT-dimensional decision variable $(\mathbf{u}(0)', \mathbf{u}(1)', \ldots, \mathbf{u}(T-1)')'$.

The Fundamental Existence Theorem of Linear Programming (Theorem 2.7) tells us that at most n of these nT elements can be nonzero. Exactly which elements are nonzero will depend on the matrix A and weighting vectors $\mathbf{w}(t)$. If the Perron root of A is larger than 1 and the weighting vectors $\mathbf{w}(t)$ are relatively constant with respect to time, then intuitively one would expect to maximize $J_T(\mathbf{x}_0)$ by letting the population grow and harvest individuals only towards the end of the time period $[0, T]$. Similarly, if the Perron root is less than 1, one would expect harvesting to occur near the beginning of the time period $[0, T]$. The precise numerical solution for any $\mathbf{x}_0$ and $\mathbf{x}(T)$, however, is easily found using the simplex algorithm for solving linear programming problems.

Thus in a linear setting, optimal harvesting theory centers around the theory of linear programming, while the character of particular applications depends on model parameters, weighting vectors, and the constraints invoked on decision and state variables. In contrast, it will become apparent in the next chapter and in the forestry chapter (Chapter 5) that unless the constraints on variables are particularly severe, the optimal solution to nonlinear harvesting problems is determined by the dynamics of the system and/or the economics embedded in the performance criterion.

2.4 STOCHASTIC THEORY

2.4.1 Introduction

Variability is more than a fact of life; it is essential to life if natural selection is to carry out its evolutionary work. Thus far we have regarded each individual in the population as being identical and subject to the same environmental forces. Both of these assumptions are obviously wrong. What can we gain by making our models more realistic at the expense of making them more complex? Only by taking cognizance of variation among individuals can we begin to address evolutionary questions (for example, see Charlesworth, 1980), and only by taking cognizance of stochastic environmental affects on a population can we begin to address questions relating to an unexpected collapse of a resource, as has happened in many of the world's most productive fisheries. By their very nature, however, models will always caricature reality. Adding more detail to models very often leads to a more obscure picture and, as often as not, a less accurate picture of reality (for example, see Hochberg *et al.*, 1986). The appropriate level of model complexity depends on the purpose of the model. In this book, we incorporate uncertainty so that risk assessment questions can be addressed in a resource management context.

Stochasticity is also a way of characterizing the impact of a number of complex environmental processes in terms of their average effect on the population. Thus from a modeling point of view, adding stochastic elements provides a technique for including factors that in a deterministic model would be too complex to include. This is especially true if we do not know the mechanistic details of how these factors influence our system but have empirical data from which we can evaluate statistically the effects of these factors on our system.

Variability between individuals implies that the values of the life-history parameters (e.g., survival and fecundity rates), in models such as equation (2.27), are actually distributed across the population. Characterizing these distributions requires sampling the population to estimate, at least, the mean and variance associated with each parameter. Environmental variability implies that parameter values for particular individuals actually vary with time according to environmental conditions. For example, survival rates may decrease in times of drought, flood, and/or fire. Not being fully able to account for this type of variability introduces *process errors* into our analysis of population development. We try to capture the essence of these errors (that is, the variability in population development) by introducing stochasticity into the model in a number of ways, as discussed below.

There is another source of uncertainty associated with *measurement errors* that are made when estimating the values of the parameters needed to run the model. This includes both population life-history parameters and the initial population state, x_0. The latter is required to begin simulating population development. Parameter estimation involves both biological and mathematical elements. The questions of how to locate individuals, estimate the age of dead individuals, or even assess whether a female has reproduced in a particular season are squarely in the realm of biology; while devising sampling strategies, estimating variance, and deciding whether parameter estimates are biased involves a number of purely mathematical issues.

The whole question of estimation is beyond the scope of this book. Biological questions relating to the problem of measuring life-history parameters are best dealt with in resource- or population-specific contexts. Insects are a case in point: it is important to know the biology of an insect to be efficient in finding individuals in the field. Some techniques for estimating population parameters are intimately tied up with the type of data that can be collected. For example, in fisheries, a technique called *virtual population analysis* has been developed to estimate cohort survival rates from data relating catch to fishing effort (Pope, 1972). The most comprehensive analysis to date of the question of parameter estimation in population harvesting is presented in Walters (1986), where questions relating to both process and measurement errors are dealt with. The problem, however, is sufficiently complicated that Walters' analysis focuses on lumped rather than stage-structured models of the detail contained in equation (2.27). (Walters' models, however, are generally nonlinear, as are the stage-structured models presented in the next chapter.)

Before we begin our discussion of stochastic issues, we need to introduce the notion of a stochastic variable X_i which represents the probable, rather than actual, number of individuals in the ith stage class. The deterministic variables x_i that satisfy the Leslie model (2.11), or the stage-structured model (2.27) can be regarded as a first cut at estimating the most probable trajectories of the stochastic variables X_i. The most we can ever know about these stochastic variables X_i are the probability distributions that characterize the path of their possible values through time. In many instances, however, we may require only the means, variances, and covariances associated with the variables X_i (we use the upper-case convention to denote stochastic variables); and, in general, it is much simpler to solve for these than it is to generate the probability distributions associated with the population process.

At an advanced level, the analysis of stochastic systems involves measure theory, a relatively abstract analytical branch of mathematical theory. We will not begin to follow this line of analysis. Rather, we will present as simply as possible those results that are required to apply stochastic analysis to resource management problems. This includes finding the means, variances, and covariances of the random variables X_i, as well as other elementary notions relating to the expectation of functions of these random variables.

We recall from probability theory (see an introductory text) that the expectation E of a function $g(X)$ of a random variable or vector of random variables X, with probability distribution $\mathcal{F}(X)$, is defined to be

$$E[g(X)] = \int_{\text{range of } X} g(X)\mathcal{F}(X)\,dX.$$

The mean $\bar{x}_i$ of the variable X_i is thus

$$\bar{x}_i = E[X_i], \qquad i = 1, \ldots, n.$$

(The more usual notation for the mean of a distribution is μ_i, while $\bar{x}_i$ is used to denote a sample estimate of μ_i. We use $\bar{x}_i$ here because it fits in with our systems-modeling notation and, below, will actually take on the interpretation of an estimate of the true population mean.)

The variance γ_{ii} (for notational convenience we use γ_{ii} rather than the more usual σ_i^2) and covariance γ_{ij} are determined by

$$\gamma_{ij} = E[X_i X_j] - \bar{x}_i \bar{x}_j, \qquad i, j = 1, \ldots, n. \tag{2.50}$$

Note by symmetry that $\gamma_{ij} = \gamma_{ji}$ for all i and j. It will also be notationally convenient to introduce the variance operator $\mathcal{V}[X]$, defined to be

$$\mathcal{V}[X] = E[X^2] - (E[X])^2. \tag{2.51}$$

Finally it will be useful for notational purposes to define the *Kronecker product* of two matrices. The Kronecker product of an $n \times m$-dimensional matrix B with an $r \times s$-dimensional matrix C, written as $B \otimes C$, is defined to be the $nr \times ms$ matrix constructed from $n \times m$ blocks of matrices of the form $b_{ij}C$; that is,

$$B \otimes C = \begin{pmatrix} b_{11}C & \cdots & b_{1m}C \\ \vdots & \ddots & \vdots \\ b_{n1}C & \cdots & b_{nm}C \end{pmatrix}, \tag{2.52}$$

where b_{ij} are the elements of B.

2.4.2 Pollard's Stochastic Model

Here we present an approach taken by Pollard (1966) in deriving an exact stochastic analogue of the Leslie matrix model (2.11) (with $s_n = 0$ in (2.12)). We will informally extend his result to the stage-structured models as well. Pollard (1966, 1973) assumed that each element s_i in the matrix L represented the probability of an individual aged i surviving to age $i + 1$, in contrast to the deterministic interpretation in which each element s_i is viewed as the proportion of individuals that survive the transition period from age i to age $i + 1$. Under this assumption, it follows from a reinterpretation of (2.3) that the random variable $X_{i+1}(t+1)$ is a binomially distributed variable conditioned on the value of the stochastic variable $X_i(t)$. On the other hand, the stochastic variables $X_i(t)$ and $X_j(t)$, $i \neq j$, represent for given t distinctly different cohorts and therefore (continuing to regard the background environment as constant) can be assumed to be independent. Under these assumptions it follows (for a proof see Pollard, 1973) that the means, variance, and covariance terms satisfy

$$\bar{x}_{i+1}(t+1) = s_i\bar{x}_i(t), \qquad i = 1,\ldots,n-1,$$
$$\gamma_{i+1\,i+1}(t+1) = s_i^2\gamma_{ii}(t) + s_i(1-s_i)\bar{x}_i(t), \qquad (2.53)$$
$$\gamma_{i+1\,j+1}(t+1) = s_is_j\gamma_{ij}(t), \qquad i \neq j,\; i,j = 1,\ldots,n.$$

In the simplest case, Pollard assumed that each female can give birth to at most one female, thereby regarding the element $\beta_i = s_0 b_i$ as the probability that a female aged i will give birth to one female that survives to the end of the transition period. For this case, it follows that (see Pollard, 1973)

$$\bar{x}_1(t+1) = \sum_{i=1}^{n} \beta_i\bar{x}_i(t),$$
$$\gamma_{11}(t+1) = \sum_{i=1}^{n} \left(\beta_i^2\gamma_{ii}(t) + \beta_i(1-\beta_i)\bar{x}_i(t)\right)$$
$$+ \sum_{i=1}^{n}\sum_{\substack{j=1 \\ j\neq i}}^{n} \beta_i\beta_j\gamma_{ij}(t), \qquad (2.54)$$
$$\gamma_{0\,j+1}(t+1) = s_j\sum_{i=1}^{n} \beta_i\gamma_{ij}(t), \qquad j = 1,\ldots,n.$$

From equations (2.53) and (2.54), we see that the vector of means, $\bar{x}$, satisfies a set of equations identical to the Leslie matrix equation (2.11), that is,

$$\bar{x}(t+1) = L\bar{x}(t). \qquad (2.55)$$

Linear stochastic systems often satisfy a *certainty equivalence principle* which, as above, refers to situations where the means of stochastic variables satisfy an analogous deterministic equation. As we shall see in the next chapter, this is not the case for nonlinear problems.

Using the Kronecker product of L with itself (see definition (2.52)) it can also be shown (Pollard, 1973) that the variance and covariance elements γ_{ij} arranged in the n^2-dimensional vector

$$\gamma = (\gamma_{11}, \ldots, \gamma_{1n}, \gamma_{21}, \ldots, \gamma_{nn})'$$

satisfy the equation

$$\gamma(t+1) = D\bar{\mathbf{x}}(t) + (L \otimes L)\gamma(t), \qquad (2.56)$$

where the $n^2 \times n$ matrix D has the form

$$D = \begin{pmatrix} \beta_1(1-\beta_1) & \beta_2(1-\beta_2) & \cdots & \beta_n(1-\beta_n) \\ \vdots & \text{n rows of zeros} & & \vdots \\ s_1(1-s_1) & 0 & \cdots & 0 \\ \vdots & \text{n rows of zeros} & & \vdots \\ 0 & s_2(1-s_2) & \cdots & 0 \\ \vdots & \vdots & \vdots & \vdots \\ \vdots & \vdots & \vdots & \vdots \\ 0 & \cdots & s_{n-1}(1-s_{n-1}) & 0 \end{pmatrix}.$$

$$(2.57)$$

The case of multiple births can be handled in the same way with one exception. Define β_{il} to be the probability that each female of age i will give birth to a number of daughters of which exactly l will survive to enter the first age class at the end of the transition interval. Define

$$\beta_i = \sum_{l=1}^{n} l\beta_{il};$$

that is, β_i is now the expected number of female offspring surviving to age 1 for a female in age class i. Pollard (1973), regarding $X_0(t+1)$ as the sum of multinomial random variables conditioned on the stochastic variables $X_i(t)$, showed that equations (2.55) and (2.56) held for this case as well, provided that the first row of the matrix D defined in (2.57) was modified so that element d_{1i} has the form

$$d_{1i} = \sum_{l=1}^{n}(l - \beta_i)\beta_{il}. \tag{2.58}$$

Clearly, the same multinomial approach can also be taken to find the exact stochastic equivalent of the stage-structured model (2.13), when the system matrix A is defined by (2.22). Again the vector of means $\bar{x}(t)$ will satisfy the same equation as the deterministic vector $x(t)$. The variance/covariance vector $\gamma(t)$ will also satisfy equation (2.56) except the matrix D will now be more complicated than is indicated by (2.57). If one assumes that whether or not an individual moves into the next stage class is independent of its past history, then, by analogy, one would conjecture that D had the following form (the actual matrix is too detailed to write out explicitly because of the limited width of this page): all zero entries except for those immediately to the right of the nonzero entries remain the same; for $i = 1, \ldots, n - 1$, the nonzero entry $s_i(1 - s_i)$ becomes $p_i s_i(1 - p_i s_i)$; the zero entries immediately to the right of these $n-1$ entries respectively become $(1 - p_{i+1})s_{i+1}(1 - (1 - p_{i+1})s_{i+1}$ (recall $p_n = 0$); the first row of entries, excepting the first entry d_{11}, remains the same or is determined by (2.58) in the case of multiple births; in the single-birth case, the first entry d_{11} is the sum of two independent binomial probabilities—that is, $d_{11} = s_0 b_1(1 - s_0 b_1) + (1 - p_1)s_1(1 - (1 - p_1)s_1)$.

2.4.3 A Stochastic Environment

If we assume, as Pollard did, that the parameters b_i, s_i, and p_i represent probabilities that are characteristic of a given population, then we are assuming variation among individuals with respect to actual fecundity, survival, and transition rates. On the other hand, if we assume that these rates change as a consequence of varying environmental factors, then we can assume—in the first instance—that these parameters are themselves stochastic variables drawn from stationary probability distributions. The analysis of such systems proceeds by setting up a system of equations that relates the stochastic variables to one another and then taking expectations to obtain equations for the means $\bar{x}$ and variance/covariance terms γ of the process.

For example, assume that the stochastic variables satisfy an equation

$$X(t+1) = AX(t), \tag{2.59}$$

where the elements of A are themselves stochastic variables. We can again take expectations to obtain an equation that has the same basic form as equations (2.55) and (2.56), that is,

$$\begin{pmatrix} \bar{x}(t+1) \\ \gamma(t+1) \end{pmatrix} = \begin{pmatrix} A & \mathbf{0} \\ D & A \otimes A \end{pmatrix} \begin{pmatrix} \bar{x}(t) \\ \gamma(t) \end{pmatrix}, \tag{2.60}$$

but the form of D is somewhat more complicated than depicted in identity (2.57). (See Pollard, 1973 for a detailed discussion on the structure of D; also see Section 3.4 in the following chapter for an example of D in the context of the Leslie matrix model.)

On the other hand, if we assume that the matrix A is deterministic, but the environment perturbs the system equations additively in each time step through the application of an *environmental noise* vector μ, we obtain the linear stochastic system

$$X(t+1) = AX(t) + \mu,$$

which has been used quite extensively in engineering applications. This model implies, however, that the effects of environmental stochasticity are independent of the size of the population, which is totally unrealistic for most population problems. In the next chapter we will consider the effects of environmental stochasticity in a more realistic nonlinear setting.

Finally we note, in models such as equation (2.55), that if the elements of A are considered as stochastic variables rather than characteristic probabilities giving rise to multinomially distributed state variables X_i, then the associated variance terms γ_{ii} are usually much larger. (See Pollard, 1973, Section 9.8, for an illustration of this point.)

2.4.4 *Harvesting Stochastic Populations*

Suppose that a resource system is modeled by the stochastic equation

$$X(t + 1) = AX(t),$$

where we assume that the parameters of the transition matrix A represent probability elements. Then, as in equation (2.38), we can assume either that a given number of individuals $u(t)$ are removed from the population in each time interval or, as in equation (2.42), that a given proportion characterized by the matrix H is removed. The latter approach implies that we need to know the actual number of individuals that survive the transition in each time period, since only then can we actually determine the number of individuals that constitute a particular proportion of the population after transition, but just prior to harvesting. This implies that we are able to exactly measure the state of the population just prior to harvesting. For some resources, such as a stand of trees, it may be possible to make an accurate assessment of the state of the population at a desired point in time (if, of course, we are willing to pay the cost). For most problems, especially fisheries, this is usually not possible.

2.5 REVIEW

In this chapter we reviewed the construction of life tables and showed how they lead to the development of an
age-structured matrix model, commonly referred to as the
Leslie model. In this model every individual proceeds in
each time step to the next age class; that is, there is a total
transition from one age class to the next. We then demonstrated how these models can be extended to incorporate
a notion of size class for which there is a partial transition from one class to the next in each time period. We
presented this model in the form of equation (2.27), that
is,

$$\mathbf{x}(t+1) = G\mathbf{x}(t) + \begin{pmatrix} s_0 x_0(t) \\ 0 \\ \vdots \\ 0 \end{pmatrix},$$

which explicitly separates the birth "input" $s_0 x_0(t)$ from
the transition process G given by (see (2.26))

$$G = \begin{pmatrix} (1-p_1)s_1 & \cdots & 0 & 0 \\ p_1 s_1 & \cdots & 0 & 0 \\ \vdots & \ddots & \vdots & \vdots \\ 0 & \cdots & (1-p_{n-1})s_{n-1} & 0 \\ 0 & \cdots & p_{n-1}s_{n-1} & s_n \end{pmatrix}.$$

There is no advantage to separating out the birth from
the transition processes in the linear model. However, this
separation is crucial in the development of the nonlinear
theory presented in the next chapter: equation (2.26) is
our point of departure for modeling nonlinear problems.

Stability properties and behavior of solutions to the
Leslie model and the above stage-structured matrix model
were discussed in the context of the general matrix model,

$$A\mathbf{x} = \lambda \mathbf{x},$$

63

especially positive linear models in which all the elements of A are nonnegative with at least one element positive. In particular, we demonstrated that the behavior of solutions to linear models depends on the eigenvalues of the matrix A. For positive models, the population will grow towards a stable stage distribution determined by the relative size of the elements of the eigenvector corresponding to the dominant eigenvalue (i.e., the eigenvalue with the largest modulus), provided this eigenvalue is real and greater than 1. The Perron-Frobenius theorem ensures that the dominant eigenvalue (also referred to as the Perron root) of a positive model is itself positive; but if it lies between 0 and 1, then the size of the population will decay to zero (even though the relative number of individuals in each stage class approaches the same stable stage distribution). Explicit equations for determining the eigenvalues and the stable stage distribution are derived for the Leslie and stage-structured models.

In this chapter we formulated the linear harvesting problems and presented the concept of the maximum sustainable yield (MSY) solution. We demonstrated that the solution to the MSY problem involves finding the proportion of individuals that must be harvested in each stage class. Since the problem is linear, the solution depends only on this proportion and not on the absolute numbers harvested. Absolute numbers can never be determined from linear models, because linear models have no equilibrium point other than zero: populations either grow without bound (Perron root is greater than 1), decay to zero (Perron root lies between 0 and 1), or, in rare singular cases, remain unchanged (Perron root equals 1). Although we mentioned that the optimal harvesting program can always involve harvesting individuals from at most two age classes, we delay proof of this result until the following chapter, where we derive an expanded result for the more general nonlinear setting.

We also discussed harvesting populations under dynamic conditions, that is, when the initial population structure does not correspond to MSY levels. Again, the linearity of the problem causes difficulties that can be solved only by making some *ad hoc* assumptions; specifically, the solution to the problem is determined by constraints that must be placed on the number of individuals harvested and/or left in the population after harvesting during one or more harvesting periods.

Harvesting nonlinear populations is an intrinsically more interesting problem, because nonlinearities often determine nontrivial equilibrium points which play a central role in the solution to the optimal harvesting problem. For this reason, most of the harvesting theory in this book is pursued in the context of nonlinear models.

In this chapter we also introduced a number of notions relating to modeling the dynamics of a stochastic population. Models can represent either inherent stochastic processes (that is, there is an inherent probability associated with the events of birth and death), or deterministic processes where the value of the model parameters are uncertain or randomly perturbed by environmental fluctuations. We demonstrated that a deterministic linear model can be reinterpreted as a model representing the mean number of individuals in each size class (certainty equivalence principle), and that a model for the variance/covariance matrix associated with the size class variables is easily generated using Kronecker products. As will become apparent in Chapters 3 and 4, an investigation of questions relating to population harvesting and resource management, using an analytical stochastic approach, often means scrapping the deterministic model and deriving a suitable stochastic model from first principles. Alternatively, we will discuss how harvesting policies can be derived using numerical simulation techniques.

CHAPTER THREE

Nonlinear Models

3.1 BACKGROUND

3.1.1 Scope

Partitioning models into linear and nonlinear categories is akin to dividing the universe into earth and non-earth. Earth is accessible but represents a confined space in an almost infinite universe. Studying it gives us some limited insights into the structure of the universe, but many of the phenomena that occur in the universe do no occur on earth. The earth is our home base and, as such, is the jumping-off point to exploring the great beyond. Before we embark on such a journey, however, we need to define our mission and prepare adequately for the voyage. The previous chapter represents that preparation. Our mission is the construction of class-structured resource management models that reflect the realities of our resource (space, food) limited world. Finally, we will remain within visual contact of our linear earth. Thus instead of beginning our discussion by replacing the linear equation (2.13) with a general class of nonlinear discrete time equations of the form

$$\mathbf{x}(t + 1) = \mathbf{f}\big(\mathbf{x}(t)\big) \qquad (3.1)$$

(where $\mathbf{f}$ is a function that maps points in R^n onto other points in R^n), we will build nonlinearities into population growth equation (2.27) leaving as much linear structure intact as we possibly can. We will start cautiously by including a single, and possibly most important, nonlinearity before going on to develop a more general theory.

As population densities increase, competition among individuals for the available resources begins to take effect, typically by increasing mortality rates, decreasing fecundity rates, and slowing down the individual growth rates. In some situations, individuals may be competing only with similarly aged individuals—for example, the larval stage of most moths and butterflies feeds on plants while the adult stage feeds on nectar. Often, however, competition is between individuals in a range of age or size classes where the intensity of competition is measured by the mass of the individuals or some other weighting factor. A weighted sum of the number of individuals in each stage class will be referred to as an *aggregation* variable. The number of newborns x_0 introduced in (2.5) is an example of such a variable. As will become apparent in our development of a nonlinear theory, if we can introduce the nonlinearities as functions of a small number of aggregation variables, then much of the linear structure can be preserved and the dimension of the nonlinearity is low with respect to dimensionality of the overall model. This facilitates both the development of theory and the numerical solution of problems.

3.1.2 Linearization and Stability

Before going on to develop a nonlinear aggregated variable theory pertaining to age- and stage-structured populations, for the sake of completeness we briefly outline a method for analyzing the stability properties of equilibrium solutions to the system of equations (3.1). An equilibrium solution (that is, equation (2.19) is satisfied) to equations (3.1) is the constant vector $\hat{\mathbf{x}}$ that satisfies the equation

$$\hat{\mathbf{x}} = \mathbf{f}(\hat{\mathbf{x}}). \tag{3.2}$$

Define a vector $\triangle\mathbf{x}(t)$ as the deviation of $\mathbf{x}(t)$ from the value $\mathbf{x}$; that is,

$$\triangle\mathbf{x}(t) = \mathbf{x}(t) - \hat{\mathbf{x}}, \qquad t = 0, 1, 2, \ldots. \tag{3.3}$$

Then using equation (3.2) and Taylor's expansion for functions of several variables, it follows that equation (3.1) can be written as

$$\triangle x(t+1) = \triangle x(t)\left(\frac{d\mathbf{f}}{d\mathbf{x}}\bigg|_{\hat{x}}\right.$$

$$\left. + \text{ terms that go to zero as } \triangle x(t) \to 0\right), \quad (3.4)$$

where $\frac{d\mathbf{f}}{d\mathbf{x}}\big|_{\hat{x}}$ is the so-called *Jacobian matrix* whose ijth element is the derivative $\frac{df_i}{dx_j}$ evaluated at $\hat{x}$.

If the linearized system is either stable (all eigenvalues lie within the unit circle 1) or unstable (at least one eigenvalue lies outside the unit circle), then $\triangle x(t)$ respectively decays to zero or grows away from zero. Since, from the Taylor expansion, the linearized approximation is valid in a neighborhood of $\hat{x}$, it follows from (3.3) that the stability properties of the linearized system reflect the behavior of the solutions to the nonlinear system in the neighborhood of the equilibrium solution $\hat{x}$. Thus we cannot conclude that the nonlinear system has a globally stable equilibrium solution when the linearized system has a globally stable equilibrium solution. In fact, the nonlinear system may have several stable solutions. Note that the only case in which the local properties of $\hat{x}$ cannot be assessed by linearization is if one or more eigenvalues of the Jacobian matrix lie on the unit circle while the remaining eigenvalues lie within the unit circle.

3.2 DENSITY-DEPENDENT RECRUITMENT

3.2.1 Density Dependence and Offspring

We begin our discussion of nonlinear problems by first considering age- rather than more complicated stage-structured populations. An age-structured population can maintain its numbers or grow only if each breeding adult is able, on average, to at least replace itself from one generation to the next. A population's ability to grow depends on both survivorship and fecundity. The form of this dependence is expressed by equation (2.35), which we recall is

$$\sum_{i=1}^{n} \frac{b_i l_i}{\lambda^i} = 1. \tag{3.5}$$

From this equation, it is clear that the Perron root λ_Δ (the real positive value of λ that satisfies equation (3.5)) increases if any of the terms $b_i l_i$ increase. In the reproductive fitness game, some species have opted for relatively large fecundity (increasing the value of positive b_i and/or decreasing the age-to-maturity by reducing the value of i for which the first positive b_i occurs). These are the so-called r-strategist species that do best in highly variable environments. Other species have invested in high survival typically at the expense of fecundity through such mechanisms as increased parental care of young. These are the so-called K-strategists (for a more detailed discussion of r- and K-strategists see, *inter alia*, Emlen, 1984).

In general, the value of the eventual population growth rate, λ_Δ, is most sensitive to s_0, the survival of the youngest age class, since s_0 is the only parameter (recalling (2.2)) to affect all the values l_i in equation (3.5). (More correctly, if i_0 is the age of reproductive maturity—whence $b_i = 0$ for $i = 1, \ldots, i_0 - 1$—then the product $l_{i_0} = s_0 \times \cdots \times s_{i_0-1}$ affects all the nonzero terms in equation (3.5), and λ_Δ is most sensitive to l_{i_0}.) In most species (including humans), before postreproductive senescence sets in, individuals are most at risk in their first year (or stage class). Furthermore, competition among adults for rearing sites (nests, dens, etc.) and/or food, or competition between large numbers of developing young for refuges and food, is likely to cause increasing mortality rates in the individual young with increasing breeding-population density. Thus nonlinearities could well be evident in s_0, the percentage mortality in the first period of a cohort's existence. In particular, s_0 will decrease in value at high densities of some aggregated population variable.

Let us return to the basic linear age-structured mortality/aging relationships defined by equations (2.3). For $i = 1$ we assume that the number of individuals born at time t surviving to age 1 is no longer the linear expression $s_0 x_0$, but is modified by a nonlinear function ψ, to become

$$x_1(t + 1) = s_0 x_0(t)\psi\big(x_0(t)\big). \qquad (3.6)$$

If the relationship between newborn individuals and one-year-olds is strictly compensatory, that is, per capita survival decreases with increasing density of newborns, then $\psi(0) = 1$ and $\psi(x_0) \geq 0$ is a decreasing function of x_0. If the remaining equations in (2.3) (that is, $i = 1, \ldots, n - 2$) are assumed to remain linear, then it follows, as in deriving equations (2.6), that (assuming $s_n = 0$ and $l_0 = 1$)

$$x_i(t) = \frac{l_i}{s_0} x_1(t - i + 1), \quad i = 1, \ldots, n - 1. \qquad (3.7)$$

By definition, x_0 is an aggregated variable representing the number of eggs laid or new births in the population (see relationship (2.5)). Thus the nonlinear function ψ reflects density-dependent competition between young for resources. The aggregated variable x_0 could also be defined as the total biomass of the population, in which case such density-dependent phenomena as adult cannibalism of young (salmon is a case in point, as discussed in Chapter 4) could be included in ψ. If we ignore other density-dependent processes under the assumption that they are much less significant than density-induced effects on first-period survival, then equations (3.7) apply.

With the introduction of the nonlinear recruit modification function ψ several equations derived in Chapter 2

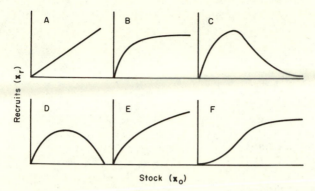

FIGURE 3.1. Nonlinear relationships between newborn individuals (x_0) and survivorship to age r (x_r); see equation (3.8). A, linear; B, Beverton and Holt; C, Ricker; D, Schaefer; E, power function ($0 < \beta < 1$); F, depensation.

need to be modified to include ψ. In particular equation (2.7) now has the form

$$x_0(t+1) = \sum_{i=1}^{n} b_i l_i x_0(t - i + 1)\psi\big(x_0(t - i + 1)\big).$$

In the fisheries literature, the function ψ introduced in equation (3.6) is referred to as a *stock-recruitment* relationship. In cases where the fish become available to the fishing gear only at age r rather than at age 1, a relationship between $x_r(t+r)$ and $x_0(t)$ is inferred from available catch and fishing effort data; that is, equation (3.6) is generalized (recalling that $\prod_{i=0}^{r-1} s_i = l_r$) to

$$x_r(t + r) = l_r x_0(t)\psi\big(x_0(t)\big). \tag{3.8}$$

Several of the most common relationships (they are not all strictly compensatory as defined above), together with names that are often attached to them, are depicted in Figure 3.1. Mathematical expressions of the nonlinear recruitment modification function ψ are (note that each of these forms involves a single parameter $\beta > 0$):

71

Beverton and Holt

$$\psi(x_0) = \frac{1}{1 + \beta x_0}; \qquad (3.9)$$

Ricker

$$\psi(x_0) = e^{-\beta x_0}; \qquad (3.10)$$

Schaefer

$$\psi(x_0) = (1 - \beta x_0); \qquad (3.11)$$

Power function

$$\psi(x_0) = x_0^{-\beta}; \qquad (3.12)$$

Depensation

$$\psi(x_0) = \frac{x_0}{1 + \beta x_0^2}. \qquad (3.13)$$

As depicted in Figure 3.1, except for the power function relationship, x_1 is bounded above for all values of x_0. Further, in the Beverton and Holt and the depensatory relationships this upper bound γs_0 is only approached as $x_0 \to \infty$. The depensatory and power functions also deviate from the standard assumption that $\psi(0) = 1$; for these two cases we respectively have $\psi(0) = 0$ and $\psi(0) = \infty$.

The following three-parameter function has also been employed in a general analysis of population harvesting (Schnute, 1985):

Deriso-Schnute

$$\psi(x_0) = (1 - \beta\gamma x_0)^{1/\gamma}. \qquad (3.14)$$

It includes as special cases (see Schnute, 1985): the linear ($\gamma = -\infty$), Beverton and Holt ($\gamma = -1$), Ricker ($\gamma = -0$), and Schaefer ($\gamma = 1$) functions.

Sometimes it is convenient to replace $l_r x_0 \psi(x_0)$ in equation (3.8) with the more compact notation $\phi(x_0) = l_r x_0 \psi(x_0)$ in which case this equation is simply written as

$$x_r(t + r) = \phi\big(x_0(t)\big), \qquad (3.15)$$

where $r \geq 1$.

3.2.2   Equilibrium Solutions

Suppose we impose equilibrium conditions (2.19) on equation (3.6) and assume there exists an equilibrium solution $\hat{x}$. Then recalling expression (2.5) for the aggregated variable x_0, it follows from equation (3.6) that the elements of $\hat{x}$ satisfy

$$\hat{x}_1 = s_0 \sum_{i=1}^{n} b_i \hat{x}_i \psi \left(\sum_{i=1}^{n} b_i \hat{x}_i \right). \qquad (3.16)$$

Equations (3.7), under equilibrium conditions, simply reduce to

$$\hat{x}_i = \frac{l_i}{s_0} \hat{x}_1, \qquad i = 1, \ldots, n-1, \qquad (3.17)$$

which can now be applied to equation (3.16) to obtain the following equation in the variable $\hat{x}_1$ alone:

$$\psi \left(\sum_{i=1}^{n} b_i l_i \hat{x}_1 / s_0 \right) = \frac{1}{\sum_{i=1}^{n} b_i l_i}. \qquad (3.18)$$

Note that if $s_n \neq 0$, it follows from equation (2.4) that, under equilibrium conditions,

$$\hat{x}_n = \frac{s_n}{1 - s_{n-1}} \hat{x}_{n-1}.$$

Thus equation (3.18) still holds if we define

$$l_n = \frac{1}{1 - s_{n-1}} \prod_{j=0}^{n-1} s_j,$$

but this modification of l_n does not apply to dynamic situations or the stability analysis discussed in the following section.

We previously defined $R_0 = \sum_{i=1}^n b_i l_i$ (recall definition (2.8)). Thus equation (3.18) is, in terms of the reproductive value parameter R_0,

$$\psi\left(R_0 \hat{x}_1 / s_0\right) = \frac{1}{R_0}, \tag{3.19}$$

or, if the function ψ is invertible (which is the case if it is strictly decreasing),

$$\hat{x}_1 = \frac{s_0}{R_0} \psi^{-1}\left(\frac{1}{R_0}\right). \tag{3.20}$$

Once the equilibrium value $\hat{x}_1$ has been calculated from (3.20), then the remaining elements $\hat{x}_i$ of the equilibrium solution $\hat{x}$ follow directly from equation (3.17). Note also from relationships (2.5), (2.8), and (3.17) that at equilibrium

$$\hat{x}_0 = \frac{R_0 \hat{x}_1}{s_0}. \tag{3.21}$$

Consider the case where ψ has the form applicable to the Deriso-Schnute relationship (3.14): that is, $\psi(x_0) = (1 - \beta\gamma x_0)^{1/\gamma}$. In this case (3.21) explicitly reduces to (after applying equation (3.20))

$$\hat{x}_0 = \frac{1 - R_0^{-\gamma}}{\beta\gamma}. \tag{3.22}$$

Since a unique biologically nontrivial equilibrium exists, by definition, only if $x_0 > 0$, then $\beta > 0$ implies that

$$\frac{1 - R_0^{-\gamma}}{\gamma} > 0. \tag{3.23}$$

It is easily shown that this inequality holds for all $\gamma \in (-\infty, \infty)$ if and only if

$$R_0 > 1. \tag{3.24}$$

74

In fisheries the quantity R_0 is sometimes referred to as the *stock value* rather than the more common designation of *reproductive value*.

It is interesting to compare inequality (3.24) with relationship (3.5) derived for the linear equivalent of our nonlinear model. If relationship (3.5) has a solution $\lambda > 1$, then inequality (3.24) is satisfied; that is, an equilibrium exists for the Deriso-Schnute model if and only if the life-table parameters b_i, l_i, $i = 1, \ldots, n$, correspond to an exponentially growing Leslie matrix model (eggs or newborn individuals have a stock value greater than 1).

Note that when $\gamma = -1$ (the Beverton and Holt case), equation (3.22) reduces to

$$\hat{x}_0 = \frac{R_0 - 1}{\beta},$$

while, when $\gamma = 0$ (the Ricker case), it can be shown that equation (3.22) reduces to

$$\hat{x}_0 = \frac{\ln R_0}{\beta}.$$

The power function relationship is easily solved to obtain

$$\hat{x}_0 = R_0^{1 + \frac{1}{\beta}}. \tag{3.25}$$

Unlike the Deriso-Schnute relationship which has a unique biologically nontrivial equilibrium solution if and only if inequality (3.24) is satisfied, it follows from equation (3.25) that the power function always yields a biologically nontrivial equilibrium solution. The reason for this is the unrealistic behavior of relationship $\psi(x_0) = x_0^{-\beta}$ (see (3.12)) at $x_0 = 0$: as the egg or newborn index x_0 approaches zero, the per capita survival rate $\psi(x_0)$ approaches infinity.

Finally, the equilibrium solution involving the depensatory relationship can be shown to satisfy a quadratic

75

equation which will have two, one, or no biologically non-trivial equilibrium solutions,

$$\hat{x}_0 = \frac{R_0 \pm \sqrt{\gamma R_0 - 4\beta}}{2\beta},$$

for R_0 respectively greater than, equal to, or less than $\frac{4\beta}{\gamma}$.

3.2.3 Stability Properties

In the first section to this chapter we outlined how the Jacobian matrix $\frac{d\mathbf{f}}{d\mathbf{x}}$ associated with equation (3.1) can be used to assess the properties of its equilibrium solutions.

The model developed above is particularly easy to linearize since the only nonlinear terms are those that pertain to the first-year survival modifier function ψ in equation (3.6), that is, the model has the form

$$\mathbf{x}(t+1) = \begin{pmatrix} 0 & \cdots & 0 & 0 \\ s_1 & \cdots & 0 & 0 \\ \vdots & \ddots & \vdots & \vdots \\ 0 & \cdots & 0 & 0 \\ 0 & \cdots & s_{n-1} & 0 \end{pmatrix} \mathbf{x}(t) + \begin{pmatrix} s_0 x_0 \psi(x_0) \\ 0 \\ \vdots \\ 0 \\ 0 \end{pmatrix}.$$

$$(3.26)$$

Recalling $x_0 = \sum_{i=1}^n b_i x_i$ (relationship (2.5)), it follows that

$$\frac{d}{dx_i}\big(s_0 x_0 \psi(x_0)\big) = k s_0 b_i, \qquad (3.27)$$

where k is the derivative of $x_0 \psi(x_0)$ evaluated at the equilibrium value $\hat{x}_0$; that is,

$$k = \frac{d}{dx_0}[x_0 \psi(x_0)]\Big|_{\hat{x}_0}. \qquad (3.28)$$

It now easily follows from (3.27) that the Jacobian matrix associated with the nonlinear system (3.25) (see equation (3.4)) has the form

$$\frac{d\mathbf{f}}{d\mathbf{x}} = \begin{pmatrix} k s_0 b_1 & \cdots & k s_0 b_{n-1} & k s_0 b_n \\ s_1 & \cdots & 0 & 0 \\ \vdots & \ddots & \vdots & \vdots \\ 0 & \cdots & 0 & 0 \\ 0 & \cdots & s_{n-1} & 0 \end{pmatrix}. \qquad (3.29)$$

Depending on the form of the function ψ, it is possible that the derivative $\frac{d\psi}{dx_0}$ has a sufficiently large negative value when evaluated at the equilibrium value $\hat{x}_0$ to make k, as defined in (3.28), negative. If k is positive, however, then the Jacobian matrix (3.29) is a Leslie matrix (see (2.12) and note that we are treating $s_n = 0$). Thus the linearized model is a Leslie model with the first-year survival rate modified by the value of k.

For $k > 0$ the Perron-Frobenius theorem discussed in the previous section applies, but for all values of k the eigenvalues satisfy an equation that is analogous to (3.5); that is,

$$1 = \sum_{i=1}^{n} \frac{k b_i l_i}{\lambda^i}, \qquad (3.30)$$

provided $\lambda \neq 0$.

It follows, as in the proof relating to Result 2.5 and from the definition of R_0 in expression (2.8), that the linearized system associated with the Jacobian (3.29) is unstable if $k R_0 > 1$ and is stable if $0 < k R_0 < 1$. Using a well-known result from matrix theory (see Reed, 1980), however, it is possible to show that the eigenvalues associated with Jacobian matrix (3.30) lie within the unit circle as long as $-1 < k R_0 < 1$. Combining these remarks and recalling previous discussions on the relationship between the stability of a nonlinear and its corresponding linearized system, the following result can be stated.

Result 3.1 *The system of nonlinear equations (3.26) is stable at an equilibrium corresponding to an egg production or new birth's level $\hat{x}_0$ when*

$$\frac{-1}{R_0} < \frac{d}{dx_0} \left[x_0 \psi(x_0) \right] \Big|_{\hat{x}_0} < \frac{1}{R_0}, \qquad (3.31)$$

and is unstable when

$$\frac{d}{dx_0} [x_0 \psi(x_0)] \Big|_{\hat{x}_0} > \frac{1}{R_0}. \qquad (3.32)$$

Note that no conclusions can be reached without further analysis if

$$\frac{d}{dx_0}[x_0\psi(x_0)]\Big|_{\hat{x}_0} < -\frac{1}{R_0}.$$

We now return to the Deriso-Schnute recruitment function (3.14). For this function $\psi(x_0) = (1 - \beta\gamma x_0)^{1/\gamma}$, where the equilibrium value $\hat{x}_0$ is given by equation (3.22). Thus it can be shown, for the Deriso-Schnute function, that

$$\frac{d}{dx_0}[x_0\psi(x_0)]\Big|_{\hat{x}_0} = R_0^{\gamma-1}\left(1 - \frac{\gamma+1}{\gamma}\left(1 - R_0^{-\gamma}\right)\right). \quad (3.33)$$

Substituting this expression in inequalities (3.31) and (3.32) it follows from Result 3.1 that the system is stable whenever

$$-R_0^{-\gamma} < 1 - \frac{\gamma+1}{\gamma}\left(1 - R_0^{-\gamma}\right) < R_0^{-\gamma},$$

that is,

$$1 + R_0^{-\gamma} > \frac{\gamma+1}{\gamma}\left(1 - R_0^{-\gamma}\right) > 1 - R_0^{-\gamma}.$$

A biologically nontrivial equilibrium solution exists for the Deriso-Schnute function if and only if inequality (3.23) is satisfied, whence it follows that inequalities (3.31) reduce to

$$\gamma\frac{1 + R_0^{-\gamma}}{1 - R_0^{-\gamma}} > \gamma + 1 > \gamma. \quad (3.34)$$

The right-hand side of this inequality is trivially true, which implies for the Deriso-Schnute function that inequality (3.32) can never hold. Thus an unstable biologically nontrivial equilibrium can arise for the Deriso-Schnute version of equation (3.26) only when the left-hand inequality in (3.34) is violated. Hence the transition from certain stability to possible instability occurs at points in the parameter space satisfying

$$\gamma + 1 = \gamma \frac{1 + R_0^{-\gamma}}{1 - R_0^{-\gamma}},$$

or, solving for R_0, when

$$R_0 = (1 + 2\gamma)^{1/\gamma}.$$

Since the left-hand side of this equation is always positive, it can be satisfied only for $\gamma \geq 0$. For the Beverton and Holt model ($\gamma = -1$), inequality (3.34) is always satisfied so that the unique nontrivial biological equilibrium, if it exists (inequality (3.24) holds), is always stable (also see Reed, 1980). The first time instabilities can arise, as a function of the value of the parameter γ increasing on $(-\infty, \infty)$, is the Ricker case when $\gamma = 0$. In this case, the right-hand side of inequality (3.34) has the value e^2. Thus we can use Result 3.1 to conclude that the Ricker form of model (3.26) has a stable nontrivial biological equilibrium whenever the inequality $1 < R_0 < e^2$ is satisfied. The equilibrium solution, however, may become unstable when $R_0 > e^2$. To draw a stronger conclusion requires additional analysis. For example, Levin and Goodyear (1980) have shown that some solutions to the Ricker model exhibit undamped oscillations with interesting periodic properties. The whole question of periodicity is rather complex and the reader is referred to Levin and Goodyear for further discussion (also see May and Oster, 1976; Guckenheimer *et al.*, 1976).

Analysis of the stability properties of equation (3.26) with respect to the remaining recruitment functions can be similarly undertaken.

3.2.4 Sustainable Harvests

So far, we have analyzed only the behavior of our age-structured nonlinear first-year survival model under un-exploited conditions. In this section we will discuss the question of exploitation in the context of proportional harvesting under equilibrium conditions. We will use the quantities h_i, introduced in the last chapter, to represent the proportion of individuals removed from the ith age class, $i = 1, \ldots, n$ at the end of each time period (see Figure 2.1). Also recall that since h_i is a proportion, its value ranges between 0 and 1 (cf. inequalities (2.41).

If we compare equation (3.26) with (2.27) then, under proportional harvesting, it follows from equation (2.45) that equilibrium solutions satisfy the equation

$$\mathbf{x} = \bigl(I - (I - H)G\bigr)^{-1} \begin{pmatrix} s_0 x_0 \psi(x_0) \\ 0 \\ \vdots \\ 0 \end{pmatrix}, \qquad (3.35)$$

where in G, as defined by (2.26), we have $p_i = 1$, $i = 1, \ldots, n - 1$ and $s_n = 0$. Note that we no longer write $\hat{\mathbf{x}}$, since the actual equilibrium value of $\mathbf{x}$ satisfying equation (3.35) is now dependent on our choice of harvesting proportions h_i, $i = 1, \ldots, n$. With these parameter values and recalling that $l_{i+1} = \prod_{j=0}^{i} s_j$ (identity (2.2)), we can easily show that the elements x_i in equation (3.35) are given by (cf. equations (2.47), $p_i = 1$, $i = 1, \ldots, n - 1$)

$$x_i = l_i \prod_{j=1}^{i} (1 - h_j) x_0 \psi(x_0) \qquad i = 1, \ldots, n. \qquad (3.36)$$

If we define the equilibrium harvest, as we did in Problem 1 in the previous chapter (see expression (2.43)), to be the weighted sum of individuals removed from each age class, then equation (2.48) modified to include $\psi(x_0)$

applies; that is, setting $p_i = 1$, $i = 1, \ldots, n-1$ and $s_n = 0$ in (2.48) we obtain

$$J = s_0 x_0 \psi(x_0) \sum_{i=1}^{n} w_i l_i h_i \prod_{j=1}^{i-1} (1 - h_j). \qquad (3.37)$$

At this point it is convenient to introduce the notation

$$\nu_i = \prod_{j=1}^{i} (1 - h_j), \qquad i = 1, \ldots, n, \qquad (3.38)$$

and define $\nu_0 = 1$. As a consequence of the inequalities $0 \leq h_i \leq 1$, $i = 1, \ldots, n$ (see (2.41)), the quantities ν_j obviously satisfy the inequalities

$$1 = \nu_0 \geq \nu_1 \geq \cdots \geq \nu_n \geq 0. \qquad (3.39)$$

The quantity ν_i also represents the proportion of individuals that escaped harvesting up to age i. Thus the ν_i are the modification that harvesting induces on the survival function l_i, and equations (3.36) can now simply be written as (cf. expression (3.17))

$$x_i = l_i \nu_i x_0 \psi(x_0) \qquad i = 1, \ldots, n. \qquad (3.40)$$

With this observation, we can directly modify a number of earlier expressions to include harvesting; for example, expression (3.20) becomes

$$x_1 = g \left(\sum_{i=1}^{n} b_i l_i \nu_i \right), \qquad (3.41)$$

where g is the one-year-old equilibrium-determining function that obviously has the form

$$g(\mu) = \frac{s_0}{\mu} \psi^{-1}(\mu), \qquad (3.42)$$

81

and

$$\mu = \sum_{i=1}^{n} b_i l_i \nu_i$$

is the *harvested stock value* that is now a variable depending on the harvest proportions h_i (cf. the definition of R_0 given by expression (2.8)). Note that $\mu \geq 0$ and also that g is defined only if the function ψ is invertible.

It now follows, using relationships (3.6), (3.38), and (3.41), that x_0 can be eliminated from expression (3.37) to obtain

$$J = g(\mu) \sum_{i=1}^{n} w_i l_i (\nu_{i-1} - \nu_i). \qquad (3.43)$$

In the parlance of fisheries management, the harvest J is expressible as the product of the total number of recruits ($g(\mu)$—individuals entering age class 1) and the average *yield per recruit* as determined by $\eta(\nu) = \sum_{i=1}^{n} w_i l_i (\nu_{i-1} - \nu_i)$, where $\nu = (\nu_1, \ldots, \nu_n)'$. Thus choosing ν to maximize J, as expressed in (3.43), can be accomplished by maximizing $\eta(\nu)$ for each harvested stock value μ in the range $[0, R_0]$ of possible stock values (as previously noted R_0 is the stock value in the absence of harvesting) and then choosing the particular stock value that maximizes J. With this observation our maximum sustainable yield policy can be stated as follows (cf. Problem 1 in the previous chapter).

Problem 3 (Nonlinear Recruitment MSY)

$$\max_{0 \leq \mu \leq \sum_{i=1}^{n} b_i l_i} g(\mu) \eta_\mu^*,$$

where η_μ^* is the solution to the linear programming problem

$$\max_{\nu \geq 0} \sum_{i=1}^{n} w_i l_i (\nu_{i-1} - \nu_i)$$

subject to the inequality constraints (3.39) and the constraint equation

$$\eta(\nu) = \sum_{i=1}^{n} w_i l_i (\nu_{i-1} - \nu_i).$$

The second part to this problem is a linear programming formulation that, by a suitable transformation of variables, can be reduced to a problem with two constraints (one is an equation and one is an inequality; see Getz, 1980a; Reed, 1980). As a consequence of this, it is possible to show that the MSY harvesting policy involves a partial or total harvest of only one age class, or a partial harvest of one age class and a total harvest of a second (of course older) age class. The bimodality of the MSY policy was first derived for the Leslie matrix form of Problem 1 (that is, $A = L$ in equations (2.43) and (2.44)) by Beddington and Taylor (1973). Reed (1980) derived this result under the assumption that the nonlinear modifier to the stock- or egg-recruitment relationship was compensatory (that is, the function ψ is invertible). Reed was also able to show that for increasing recruitment functions, such as the Beverton and Holt and power functions (expressions (3.9) and (3.10), respectively), the bimodal MSY policy involved no harvesting until the age i at which the survival discounted biomass $w_i l_i$ is maximized. Getz (1980a) showed that the bimodal harvesting result holds irrespective of the form of the nonlinear recruitment modifier function ψ. The approach we used to obtain this result is employed below to obtain a more general multimodal harvesting result for the stage-structured model with aggregated nonlinearities.

3.3 NONLINEAR STAGE-CLASS MODELS

3.3.1 *Model Structure*

We begin our general nonlinear stage-class modeling

treatment by recalling the linear stage model (2.27) and the nonlinear recruitment survival extension depicted in equation (3.6). The ideas behind these two sets of equations can be combined to obtain the system

$$\mathbf{x}(t+1) = G\mathbf{x}(t) + \begin{pmatrix} s_0 x_0(t)\psi\big(x_0(t)\big) \\ 0 \\ \vdots \\ 0 \end{pmatrix}. \qquad (3.44)$$

This equation, however, represents a population closed to migration, that is, individuals can enter only at birth and leave when they die. Migration is easily included by replacing the birth input vector $\big(s_0 x_0(t)\psi\left(x_0(t)\right), 0, \ldots, 0\big)'$ with the general input vector $\boldsymbol{\phi}(t) = (\phi_1, \ldots, \phi_n)$, where ϕ_i is the net number of individuals entering or exiting (ϕ_i is negative) the ith stage class at time t from sources outside the population including births (cf. equation (3.15) with $r = 1$). The element ϕ_1 is just the number of births. In the fisheries literature, ϕ_1 is sometimes referred to as the number of new recruits. The elements ϕ_i, $i > 1$ are attributable to migration processes both into and out of the population. We can also extend our treatment to include the harvesting term $\mathbf{u}(t)$, as was done in equation (2.39), so that the system (3.44) now has the general form

$$\mathbf{x}(t+1) = G\mathbf{x}(t) - \mathbf{u}(t) + \boldsymbol{\phi} \qquad t = 0, 1, 2, \ldots. \qquad (3.45)$$

Recall that $G = PS$, where the survival matrix S and the transition matrix P are respectively defined by (2.24) and (2.25) (see expression (2.26)). Also recall, in the case of harvesting, that the state variable $\mathbf{x}(t)$ has the interpretation of the residual state of the population after harvesting at the end of the transition period $[t, t+1]$.

As long as the elements of S and P are constant, equation (3.45) is a linear system with a possible nonlinear input vector $\boldsymbol{\phi}$. In general, however, transition and survivorship rates are density dependent so that these elements are nonlinear functions of the state vector $\mathbf{x}$. Here we develop a model where the dependence is on aggregations of the variables x_i (that is, weighted summations of these variables across i) rather than on the individual values of x_i themselves. We begin by introducing an $n \times m$ aggregation matrix Θ (with elements θ_{ij}). Then we define an m-dimensional aggregation vector $\mathbf{y} = (y_1, \ldots, y_m)'$ as a linear transformation of the variables $\mathbf{x}$; that is,

$$\mathbf{y} = \Theta \mathbf{x}. \tag{3.46}$$

We now assume that the parameters p_i, s_i, and ϕ_i depend on $\mathbf{y}$; that is, $\phi = \phi(\mathbf{y})$ and the elements of the matrix $G(\mathbf{y})$ have the form (cf. (2.26)) and recall that $p_n = 0$)

$$g_{ij}(\mathbf{y}) = \begin{cases} \big(1 - p_i(\mathbf{y})\big)s_i(\mathbf{y}), & j = i, \, i = 1, \ldots, n; \\ p_{i-1}(\mathbf{y})s_{i-1}(\mathbf{y}), & j = i - 1, \, i = 2, \ldots, n; \\ 0, & \text{otherwise.} \end{cases} \tag{3.47}$$

Equation (3.45) can now be written as

$$\mathbf{x}(t + 1) = G\big(\mathbf{y}(t)\big)\mathbf{x}(t) - \mathbf{u}(t) + \phi\big(\mathbf{y}(t)\big), \tag{3.48}$$

where the dependence of the population parameters on $\mathbf{y}$ is now explicitly indicated.

3.3.2 Equilibrium Theory

A first step towards understanding the properties of equation (3.48) is to analyze the system under equilibrium conditions $\mathbf{x}(t) = \mathbf{x}$ for all t. In linear systems we saw that unless the system matrix is nonsingular, the only equilibrium point is the trivial solution $\mathbf{x} = 0$. In nonlinear systems the situation is much more interesting as a number of equilibria might exist. Finding these equilibria and characterizing their stability properties is the essential problem of equilibrium analysis.

We begin this analysis by setting $\mathbf{u}(t)$ to a given constant value $\hat{\mathbf{u}}$ for all $t = 0, 1, 2, \ldots$. Of course an equilibrium

85

analysis in the absence of harvesting corresponds to the special case $\hat{\mathbf{u}} = 0$. It follows from the equilibrium condition and equation (3.48) that the corresponding values $\hat{\mathbf{x}}$ and $\hat{\mathbf{y}}$ satisfy the equation

$$\hat{\mathbf{x}} = \left(I - G(\hat{\mathbf{y}})\right)^{-1}\left(\boldsymbol{\phi}(\hat{\mathbf{y}}) - \hat{\mathbf{u}}\right), \tag{3.49}$$

or from equation (3.46)

$$\hat{\mathbf{y}} = \Theta\left(I - G(\hat{\mathbf{y}})\right)^{-1}\left(\boldsymbol{\phi}(\hat{\mathbf{y}}) - \hat{\mathbf{u}}\right). \tag{3.50}$$

Result 2.4 (with $\lambda_\Delta = 1$) can be applied to the problem of inverting the matrix $\left(I - G(\hat{\mathbf{y}})\right)$ in equation (3.49) and the terms multiplied out to obtain, for $i = 1, \ldots, n$,

$$\hat{x}_i = \frac{1}{p_i(\hat{\mathbf{y}})s_i(\hat{\mathbf{y}})} \sum_{j=1}^{i}(\phi_j(\hat{\mathbf{y}}) - \hat{u}_j) \prod_{r=j}^{i} \frac{p_r(\hat{\mathbf{y}})s_r(\hat{\mathbf{y}})}{1 - \left(1 - p_r(\hat{\mathbf{y}})\right)s_r(\hat{\mathbf{y}})}. \tag{3.51}$$

Note that the above are a system of equilibrium "balance equations" in the number of individuals in each stage class. The product term for the ith stage class determines the proportion of individuals that make it through from stage classes j to i, while $\phi_j(\hat{\mathbf{y}})$ and $\hat{u}_j$ are the number of individuals added to and subtracted from stage class j, $j = 1, \ldots, i$. Finally, it follows from equation (3.48) for $k = 1, \ldots, m$ that

$$\hat{y}_k = \sum_{i=1}^{n}\left(\frac{\theta_{ik}}{p_i(\hat{\mathbf{y}})s_i(\hat{\mathbf{y}})}\right. \\ \left. \times \sum_{j=1}^{i}(\phi_j(\hat{\mathbf{y}}) - \hat{u}_j)\prod_{r=j}^{i}\frac{p_r(\hat{\mathbf{y}})s_r(\hat{\mathbf{y}})}{1 - \left(1 - p_r(\hat{\mathbf{y}})\right)s_r(\hat{\mathbf{y}})}\right). \tag{3.52}$$

The system of equations (3.52) are implicit in $\hat{\mathbf{y}}$. For particular choices of $\hat{\mathbf{u}}$, solutions $\hat{\mathbf{y}}$ may not be unique or satisfy $\hat{\mathbf{y}} \geq 0$. In general terms, solutions to equation

(3.52) correspond to fixed points of the mapping (see the right-hand side of equation (3.50)),

$$\mathbf{y} = F_{\mathbf{u}}(\mathbf{y}) \equiv \Theta \left(I - G(\mathbf{y}) \right)^{-1} \left(\phi(\mathbf{y}) - \mathbf{u} \right). \tag{3.53}$$

For simplicity we will use the notation $\hat{p}_i = p_i(\hat{\mathbf{y}})$ and similarly define $\hat{s}_i$ and $\hat{\phi}_i$. If the equilibrium solution is to be biologically meaningful, then it follows that the elements of the control and residual state vector must satisfy

$$\left. \begin{array}{c} \hat{u}_i \geq 0 \\ \hat{x}_i \geq 0 \end{array} \right\} \quad i = 1, \dots, n. \tag{3.54}$$

From equation (3.51), this implies for $i = 1, \dots, n$ that

$$\sum_{j=1}^{i} \hat{\phi}_j \prod_{r=j}^{i} \frac{\hat{p}_r \hat{s}_r}{1 - \left(1 - \hat{p}_r \right) \hat{s}_r} \geq \sum_{j=1}^{i} \hat{u}_j \prod_{r=j}^{i} \frac{\hat{p}_r \hat{s}_r}{1 - \left(1 - \hat{p}_r \right) \hat{s}_r} \geq 0. \tag{3.55}$$

Note that

$$\prod_{r=j}^{i} \frac{\hat{p}_r \hat{s}_r}{1 - \left(1 - \hat{p}_r \right) \hat{s}_r} \geq 0$$

for all i, and $j \leq i$, since p_r and $s_r \in [0, 1]$ for all r.

3.3.3 Linearization and Stability

The stability properties of an equilibrium solution to a nonlinear system of equations can be analyzed by linearizing the system around that equilibrium solution (that is, taking the first terms in the Taylor expansion of the nonlinear model evaluated at the equilibrium values) and analyzing the eigenvalues of the derived linear system matrix. Thus the stability properties of $\hat{\mathbf{x}}$ can be analyzed by expanding equation (3.48) around the equilibrium solution pair $(\hat{\mathbf{x}}, \hat{\mathbf{u}})$. To facilitate the linearization process, define

$$\triangle \mathbf{x}(t) = \mathbf{x}(t) - \hat{\mathbf{x}} \quad \text{and} \quad \triangle \mathbf{u}(t) = \mathbf{u}(t) - \hat{\mathbf{u}}. \tag{3.56}$$

Since

$$\frac{\partial y_k}{\partial x_i} = \theta_{ki}, \qquad i = 1, \ldots, n, \quad k = 1, \ldots, m,$$

the first terms in the Taylor expansion of the nonlinear functions in equation (3.48) yield the linearized system

$$\triangle \mathbf{x}(t+1) = (\hat{G} + \hat{\Phi})\triangle \mathbf{x}(t) - \triangle \mathbf{u}(t) + \hat{K}\triangle \mathbf{x}(t), \qquad (3.57)$$

where $\hat{G}$, $\hat{\Phi}$, and $\hat{K}$ are the matrices defined by $\hat{G} = G(\hat{\mathbf{y}})$,

$$(\hat{\Phi})_{ij} = \sum_{r=1}^{n} \frac{\partial \phi_i(\hat{\mathbf{y}})}{\partial y_r} a_{rj}, \qquad (3.58)$$

and from the definition of G that

$$(\hat{K})_{ij} = \sum_{r=1}^{n} a_{rj} \left(\frac{\partial g_{i\,i-1}(\hat{\mathbf{y}})}{\partial y_r} \hat{x}_{i-1} + \frac{\partial g_{ii}(\hat{\mathbf{y}})}{\partial y_r} \hat{x}_i \right). \qquad (3.59)$$

Comparing equation (3.57) with equation (2.38) it is apparent that the matrix $\hat{G} + \hat{\Phi} + \hat{K}$ in the linearized system is equivalent to the matrix A in the linear system. Thus the stability properties of the linearized system are determined by the eigenvalues of the matrix $\hat{G} + \hat{\Phi} + \hat{K}$. If the linearized system is either stable (all eigenvalues must lie within the unit circle in the complex plain) or unstable (at least one eigenvalue lies outside the unit circle) then $\triangle \mathbf{x}(t)$ respectively decays to zero or grows away from zero. Since the linearized approximation is valid in a neighborhood of $(\hat{\mathbf{x}}, \hat{\mathbf{u}})$, it follows from (3.56) that the stability properties of the linearized system reflect the behavior of the solutions to the nonlinear system in a neighborhood of $(\hat{\mathbf{x}}, \hat{\mathbf{u}})$. The only case that cannot be assessed by linearization is if one or more of the eigenvalues lie on the circumference of the unit circle while the others lie within.

Note that equation (3.48) subject to equilibrium condition (2.19) can be used to express $\mathbf{u}$ in terms of $\mathbf{x}$ and $\mathbf{y}$; that is,

$$\mathbf{u} = \big(G(\mathbf{y}) - I\big)\mathbf{x} - \phi(\mathbf{y}). \qquad (3.60)$$

A linearization procedure can also be followed for the proportional harvesting case in which the term $\mathbf{u}$ is replaced with $H\mathbf{x}$ in equation (3.48) (recall that H is the diagonal matrix with proportional harvesting diagonal elements $0 \leq h_i \leq 1$). This equation now has the form

$$\mathbf{x}(t+1) = \big[G\left(\mathbf{y}(t)\right) - H(t)\big]\mathbf{x}(t) + \phi\left(\mathbf{y}(t)\right),$$

where, for linearization, we consider perturbations

$$\triangle\mathbf{x}(t) = \mathbf{x}(t) - \hat{\mathbf{x}} \quad \text{and} \quad \triangle h_i(t) = h_i(t) - \hat{h}_i, \quad i = 1, \ldots, n.$$

The Jacobian matrix for this system is a little different from the Jacobian matrix for equation (3.48) and so the stability properties of the resource under the two kinds of harvesting policies are different. We will not demonstrate this here, but it is known (for example, see Clark, 1976) that proportional harvesting corresponding to MSY levels can be stable when absolute harvesting corresponding to the same MSY levels is not. Intuitively, the reason for this is that under proportional harvesting, the actual amount harvested decreases as the population level (all the elements of $\mathbf{x}$) decreases. This follows since the harvesting level depends on $\mathbf{x}$ (that is, it is given by $H\mathbf{x}$). Absolute harvesting does not have this *feedback* quality.

3.4   GENERAL HARVESTING THEORY

3.4.1   Sustainable Yields

We previously discussed the problem of maximizing sustainable yields for age-structured models with at most a nonlinear first-year survival function. A generalization of the bimodal harvesting result that is applicable to such systems can be obtained. In this subsection we will state and prove this result, but the proof can be omitted by those readers not interested in the details. As in the previous sections (cf. expression (2.40)), we define the value of the harvest to be the weighted sum of the number of individuals removed from each stage class; that is, under equilibrium conditions the value of the harvest in each period is

$$J(\mathbf{u}) = \mathbf{w}'\mathbf{u},$$

where we recall that $\mathbf{w} \in R^n$ is a vector of weighting coefficients. Note that w_i is the net return from harvesting an individual in the ith stage class.

Under equilibrium conditions, we have seen that particular choices of $\mathbf{u}$ constrain $\mathbf{x}$, and hence (through equation (3.46)) constant $\mathbf{y}$, to satisfy equation (3.60). Thus recalling inequalities (3.54) and (3.55) we can state the equilibrium harvesting problem as:

Problem 4 (Nonlinear Stage-Structure MSY)

$$\max_{\mathbf{y} \in R^{+m},\ \mathbf{u} \in R^{+n}} \mathbf{w}'\mathbf{u}$$

subject to equation (3.53) and inequalities

$$\sum_{j=1}^{i} \left(\prod_{r=j}^{i} \frac{p_r(\mathbf{y})s_r(\mathbf{y})}{1 - [1 - p_r(\mathbf{y})]s_r(\mathbf{y})} \right) u_j$$

$$\leq \sum_{j=1}^{i} \left(\prod_{r=j}^{i} \frac{p_r(\mathbf{y})s_r(\mathbf{y})}{1 - [1 - p_r(\mathbf{y})]s_r(\mathbf{y})} \right) \phi_j(\mathbf{y}) \quad (3.61)$$

holding for $i = 1, \ldots, n$.

Although this problem does not specify the form of the nonlinearities in the m-dimensional vector $\mathbf{y}$, it is possible to prove the following result.

Theorem 3.2 *Suppose that at most k elements of the input vector ϕ are not identically equal to zero. If the MSY problem has an optimal solution, then it has an optimal solution $\mathbf{u}^0$ in which at most $m + k$ of its elements are nonzero.*

Proof. Conceptually, the MSY problem can be separated into a linear programming problem parameterized by $\mathbf{y}$, and the nonlinear problem of finding the best solution in the parameterized solution set. The linear programming problem, for fixed $\mathbf{y} \in R^{+m}$, is to find a solution $\mathbf{u_y}$ that maximizes $J(\mathbf{u})$ over all $\mathbf{u} \in R^{+n}$ subject to equation (3.53) and inequality (3.61). The nonlinear problem is to find a solution $\mathbf{y}^0$ that maximizes the value $J(\mathbf{u_y})$ over all $\mathbf{y} \in R^{+m}$. Hence if we can show for any $\mathbf{y} \in R^{+m}$ that the existence of a solution to the linear programming problem implies the existence of a solution that has at most $m + k$ nonzero elements, then the theorem is proven.

Suppose inequality (3.61) holds for given $\mathbf{y} \in R^{+m}$ and some i such that $\phi_i(\mathbf{y}) \equiv 0$. It follows from inequality (3.61) and the nonnegativity of the elements of $\mathbf{u}$ that

$$\sum_{j=1}^{i-1} u_j \prod_{r=j}^{i-1} \frac{p_r(\mathbf{y})s_r(\mathbf{y})}{1 - \left(1 - p_r(\mathbf{y})\right)s_r(\mathbf{y})}$$

$$\leq \frac{1 - [1 - p_r(\mathbf{y})]s_r(\mathbf{y})}{p_r(\mathbf{y})s_r(\mathbf{y})} \sum_{j=1}^{i} u_j \prod_{r=j}^{i} \frac{p_r(\mathbf{y})s_r(\mathbf{y})}{1 - [1 - p_r(\mathbf{y})]s_r(\mathbf{y})};$$

$$\leq \frac{1 - [1 - p_r(\mathbf{y})]s_r(\mathbf{y})}{p_r(\mathbf{y})s_r(\mathbf{y})} \sum_{j=1}^{i} \phi_j \prod_{r=j}^{i} \frac{p_r(\mathbf{y})s_r(\mathbf{y})}{1 - [1 - p_r(\mathbf{y})]s_r(\mathbf{y})}$$

since it is assumed that (3.61) holds for i;

$$\leq \frac{1 - [1 - p_r(\mathbf{y})]s_r(\mathbf{y})}{p_r(\mathbf{y})s_r(\mathbf{y})} \sum_{j=1}^{i-1} \phi_j \prod_{r=j}^{i} \frac{p_r(\mathbf{y})s_r(\mathbf{y})}{1 - [1 - p_r(\mathbf{y})]s_r(\mathbf{y})}$$

since $\phi_i = 0$;

$$\leq \sum_{j=1}^{i-1} \phi_j \prod_{r=j}^{i-1} \frac{p_r(\mathbf{y})s_r(\mathbf{y})}{1 - [1 - p_r(\mathbf{y})]s_r(\mathbf{y})}$$

canceling terms; that is, inequality (3.61) holds for $i - 1$. Thus inequality (3.61) holding for $i = n$ implies that this inequality holds for $i = n - 1$, $n - 2, \ldots$, $n - \ell$, where $n - \ell$ is the last element in $\phi(\mathbf{y})$ that is not identically zero. Similarly, inequality (3.61) holding for $i = n - \ell - 1$ implies that the inequalities below it hold until the second last nonidentically zero element of $\phi(\mathbf{y})$ is encountered. In this way, all but k inequalities in (3.61) are redundant. Thus the linear programming problem can be reduced to a formulation that has m equality constraints given by equation (3.53) and k of the inequality constraints in (3.61) that correspond to the nonidentically zero elements of $\phi(\mathbf{y})$. By the fundamental theorem of linear programming, if a solution exists then a basic solution exists (that is, a solution having at most $m + k$ nonzero elements), thereby proving the theorem. □

This generalizes the MSY bimodal harvesting policy result presented in the previous section. Note that (3.44) is a special case of equation (3.48) with $m = 1$ and $k = 1$.

Since resources do not serendipitously start out at MSY conditions and, further, stochastic events continually displace resources from equilibrium conditions, MSY solutions are not in themselves the most appropriate policies for managing resources. Rather, as discussed in later sections, they provide a reference point for the design of dynamic harvesting policies, preferably feedback or adaptive policies, that can respond to unforeseen changes in the state of the resource.

3.4.2 Dynamic Yield Problem

In this section, we consider the optimal yield problem in a nonequilibrium setting. A central question that arises when trying to relate equilibrium strategies to dynamic situations concerns choosing a value T for the length of the management planning horizon. The trade-off between short- and long-term gains (solutions for small versus large T) is an essential part of the management problem. Short-term gains are more attractive to individuals even in regulated resources because of the uncertainty associated with the future; one feels more confident having money in the bank than being guaranteed a share in an uncertain resource. Hence resources tend to be inefficiently exploited in the long run (Fisher, 1981). The interests of the society at large are less myopic and, hence, regulatory agencies have been created to ensure the long-term availability of the resource. This raises the question, addressed in the next section, of how to select a suitable value for T, and how to deal with boundary conditions that arise when planning over a finite time horizon. The first problem that arises is the question of choosing the length of a planning horizon.

As in Chapter 2 (recall expression (2.40)), define the value of the harvest of $[0, T]$ to be

$$J_T(\mathbf{x}_0) = \sum_{t=0}^{T-1} \mathbf{w}' \mathbf{u}(t). \qquad (3.62)$$

Note that the value of J_T depends not only on our choice of $\mathbf{u}(0), \ldots, \mathbf{u}(T-1)$, but also on $\mathbf{x}_0$, as indicated by J_T's functional argument. To avoid the problem of allowing the resource to be overexploited at time T, constrain

$$\mathbf{x}(T) = \mathbf{x}^o. \qquad (3.63)$$

Now consider the problem of selecting $\mathbf{u}(0), \ldots, \mathbf{u}(T-1)$ to maximize $J_T(\mathbf{x}_0)$ subject to equation (3.48), $t = 0, \ldots, T-1$, and final time constraint relationship (3.63). The first question that arises is a controllability question; that is, can the system be driven from $\mathbf{x}_0$ to $\mathbf{x}^o$ in T time steps? For linear systems a controllability theory exists that is not directly extendible to nonlinear systems. We can, however, approach the problem from another direction by defining a family of sets X^j, in terms of equation (3.48), the constraints

$$\left. \begin{array}{l} \hat{u}_i(t) \geq 0 \\ \hat{x}_i(t) \geq 0 \end{array} \right\} \quad i = 1, \ldots, n, \quad t = 0, 1, 2, \ldots, \qquad (3.64)$$

and the point $\mathbf{x}^o$ which we assume is the unique solution to the MSY problem defined in the previous section. Specifically, we define

$$X^j = \{\mathbf{x} \in R^{+n} \text{ that can be driven to } \mathbf{x}^o \text{ in } j \text{ time steps}\}.$$

Clearly $X^0 = \mathbf{x}^o \subseteq X^1 \subseteq X^2 \subseteq \cdots$, and if $\mathbf{x}_0 \in X^j$ but $\mathbf{x}_0 \notin X^{j-1}$ then j is the minimum number of time steps in which $\mathbf{x}_0$ can be driven to $\mathbf{x}^o$.

Results can often be derived for the optimal harvesting of a dynamic population problem, or for that matter any optimal control problem, by making use of the *principle of optimality*. For discrete time systems this principle can be stated as follows. If the solution $\tilde{\mathbf{x}}(t)$, $t = 0, \ldots, T$, is optimal for a problem defined on the interval $[0, T]$, then the solution $\tilde{\mathbf{x}}(\mathbf{t})$, $t = k + 1, \ldots, T$, is optimal for the same problem defined on the interval $[k, T]$ with specified initial condition $\mathbf{x}(k) = \tilde{\mathbf{x}}(k)$. This principle also leads directly to the formulation of a set of recursive equations that constitute the method of *dynamic programming* (for a presentation of this topic the reader is referred to Intriligator, 1971, or Luenberger, 1979).

We use the principle of optimality to prove the following theorem which determines the relationship between the MSY solution and the solution to the dynamic optimal harvesting problem.

Theorem 3.3 *If the set X^1 is convex and $\mathbf{x}(t)$ is constrained to lie in X^1 for $t = 0, \ldots, T - 1$, then the solution that maximizes expression (3.62) subject to the dynamic equation (3.48), the*

94

endpoint condition (3.63), and the constraints (3.64) drives $\mathbf{x}_0$
to $\mathbf{x}^o$ *in the first time step and keeps it there for all* $t = 1, \ldots, T$;
that is, if the optimal trajectory is denoted by $\tilde{\mathbf{x}}(t)$, *then*

$$\tilde{\mathbf{x}}(t) = \mathbf{x}^o, \qquad t = 1, \ldots, T.$$

Proof. Suppose $\mathbf{x}(j-2) = \mathbf{x}^1 \in X^1$, $\mathbf{x}(j-1)$ is constrained
to belong to X^1 and $\mathbf{x}(j) = \mathbf{x}^o$. Consider the two-time-step
problem

$$\max_{\mathbf{u}(j-2),\ \mathbf{u}(j-1)} W(\mathbf{x}(j-1)) = \mathbf{w}'[\mathbf{u}(j-2) + \mathbf{u}(j-1)], \quad (3.65)$$

subject to equation (3.48) holding for $t = j-2, j-1$; note
that $W = W\left(\mathbf{x}(j-1)\right)$ since $\mathbf{x}(j-1)$ is the only state in this
problem that can vary. Since X^1 is convex, all $\mathbf{x}(j-1) \in X^1$
can be represented by

$$\mathbf{x}(j-1) = (1-\varepsilon)\mathbf{x}^o + \varepsilon\mathbf{x}, \qquad (3.66)$$

for some $\mathbf{x} \in (X^1 - \{\mathbf{x}^o\})$ and some $\varepsilon \in [0,1]$. From equation (3.48) it follows that $\mathbf{u}(j-2)$ and $\mathbf{u}(j-1)$ can be solved
using this equation to obtain

$$\mathbf{u}(j-2) = G(\mathbf{y}^1)\mathbf{x}^1 + \mathbf{f}(\mathbf{y}^1) - (1-\varepsilon)\mathbf{x}^o - \varepsilon\mathbf{x},$$
$$\mathbf{u}(j-1) = G(\mathbf{y}^\varepsilon)\left[(1-\varepsilon)\mathbf{x}^o + \varepsilon\mathbf{x}\right] + \mathbf{f}(\mathbf{y}^\varepsilon) - \mathbf{x}^o, \quad (3.67)$$

where $\mathbf{y}^1 = A\mathbf{x}^1$ and $\mathbf{y}^\varepsilon = A[(1-\varepsilon)\mathbf{x}^o + \varepsilon\mathbf{x}]$. Thus from
equations (3.66) and (3.67) it follows that $W(\mathbf{x}(j-1))$ in
expression (3.65) can be considered to be a function $\omega(\varepsilon)$
with $\omega(0) = W(\mathbf{x}^o)$ and $\omega(1) = W(\mathbf{x})$ for all $\mathbf{x} \in (X^1 - \{\mathbf{x}^o\})$.
Hence if it can be shown that $\omega(0) \geq \omega(1)$, then it follows
that the optimal solution corresponds to $\varepsilon = 0$; that is,
from identity (3.66), the optimal value for $\mathbf{x}(j-1)$ is $\tilde{\mathbf{x}}(j-1) = \mathbf{x}^o$.

From equations (3.65) and (3.67) it follows that $\omega(0) - \omega(1)$ can be reduced to

$$\omega(0) - \omega(1) = \mathbf{w}'\Big(\big[G(\mathbf{y}^o) - I \big] \mathbf{x}^o + \mathbf{f}(\mathbf{y}^o)$$
$$- \big[G(\mathbf{y}) - I \big] \mathbf{x} - \mathbf{f}(\mathbf{y}) \Big).$$

But the right-hand side of this equation is nonnegative because $\mathbf{x}^o$ maximizes $\mathbf{w}'\mathbf{u}$ subject to equation (3.60) (see the MSY problem discussed in the previous section). Hence $\tilde{\mathbf{x}}(j - 1) = \mathbf{x}^o$, and the theorem follows by the principle of optimality and repeated iteration of this result for $j = T, T - 1, \ldots, 2$ (that is, using dynamic programming). $\qquad\square$

Under the assumptions of Theorem 3.3, the MSY solution is an integral part of the solution to the dynamic problem. It is not clear, however, how much the assumptions can be weakened and a comparable theorem still hold. For example, if $\mathbf{x}_0 \in X^j$ but $\mathbf{x}_0 \notin X^{j-1}$, then under what conditions would it be optimal to reach $\mathbf{x}^o$ in j time steps? Constraining $\mathbf{x}(t)$, in Theorem 3.3, to lie in X^1 for all t may be more restrictive than necessary. The above theorem also has limited application because of the linear form of the value function J_T given by equation (3.62), but serves to indicate that extremal equilibrium solutions may often play a central role in solutions to dynamic problems.

3.4.3 Discounted Rent and a Maximum Principle

So far we have considered only the problem of maximizing the gross value of the yield, that is, maximizing criterion (3.62). The harvests from some resources have an intrinsic value to society so that their dollar value can be expected to rise when they become scarce. In such circumstances, solving the long-term yield maximization problem may make more sense than trying to convert the yield to its dollar value and then maximizing net profit. In general, however, we can formulate a revenue maximization problem if we know how to convert the resource to its dollar value and are able to express harvesting costs as a function of the state of the resource and the harvesting level.

Let $R(\mathbf{x}(t), \mathbf{u}(t))$ denote the net revenue obtained during harvesting period $(t+1)$. We also introduce a discount factor δ to take cognizance of the fact that a dollar invested today will compound in value in each time period at the real (that is inflationary adjusted) interest rate r—specifically, $\delta = 1/(1 + r)$ (see Clark, 1985). If we plan to exploit a resource over a time period $[0, T]$, then at the end of the time period the resource has some value which we will represent by the function $L(\mathbf{x}(T), T)$. A suitable form for this function will be discussed later. Then a more general form of the value function (3.62) is

$$J_T(\mathbf{x}_0) = \sum_{t=0}^{T-1} \delta^t R\left(\mathbf{x}(t), \mathbf{u}(t)\right) + L\left(\mathbf{x}(T), T\right). \qquad (3.68)$$

Note that a value of $\delta = 1$ for the discount factor corresponds to the case where the interest rate is zero.

Before we go on to discuss the problem of maximizing the objective function (3.68) in the context of the stage-structured harvesting model (3.48), we need to introduce a set of equations that can be used to compute optimal solutions to general nonlinear problems. These conditions are often referred to as Pontryagin's Maximum Principle. Here we will state a discrete time version of this principle for systems satisfying the general nonlinear equation (extending (3.1) to include harvesting),

$$\mathbf{x}(t + 1) = \mathbf{f}\left(\mathbf{x}(t), \mathbf{u}(t)\right), \qquad (3.69)$$

for which equation (3.48) is a special case. We also assume that the initial condition

$$\mathbf{x}(0) = \mathbf{x}_0 \qquad (3.70)$$

97

is specified and that the final state $x(T)$ is either specified or constrained by a k-dimensional vector function ς, that is, by a set of relationships

$$\varsigma_i\big(\mathbf{x}(T)\big) = 0, \qquad i = 1, \ldots, k. \tag{3.71}$$

This result will also be stated in terms of a control constraint set $U \subset R^m$ where, for generality, we allow $\mathbf{u}$ to be an m-dimensional vector (usually $m \le n$). Typically the set U has the form

$$U = \big\{ \mathbf{u} \epsilon R^m \mid 0 \le u_j \le \overline{u}_j, \ j = 1, \ldots, m \big\}, \tag{3.72}$$

where $\overline{u}_j$ is the maximum number of individuals that can be harvested in a single time period.

Result 3.4 (Discrete Maximum Principle) *The solution set*

$$\Big\{ \big(\tilde{\mathbf{x}}(t), \tilde{\mathbf{u}}(t)\big) \Big| t = 0, \ldots, T \Big\}$$

maximizes (3.68) subject to equations (3.69), initial conditions (3.70), and the constraints (3.72) and (3.71) only if there exists a current value costate variable[1] $\boldsymbol{\xi}(t) \in R^n$, a multiplier $\boldsymbol{\eta}$, and current value Hamiltonian

$$\mathcal{H}(\boldsymbol{\xi}(t+1), \mathbf{x}(t), \mathbf{u}(t), t) = \boldsymbol{\lambda}'(t+1)\,\delta\,\mathbf{f}\big(\mathbf{x}(t), \mathbf{u}(t)\big) \\ + R\big(\mathbf{x}(t), \mathbf{u}(t)\big) \tag{3.73}$$

such that for $t = 0, \ldots, T-1$

$$\boldsymbol{\xi}'(t) = \boldsymbol{\xi}'(t+1)\,\delta\,\frac{\partial \mathbf{f}}{\partial \mathbf{x}}\big(\tilde{\mathbf{x}}(t), \tilde{\mathbf{u}}(t)\big) + \frac{\partial R}{\partial \mathbf{x}}\big(\tilde{\mathbf{x}}(t), \tilde{\mathbf{u}}(t)\big) \tag{3.74}$$

and

$$\mathcal{H}\big(\boldsymbol{\xi}(t+1), \tilde{\mathbf{x}}(t), \tilde{\mathbf{u}}(t), t\big) = \max_{\mathbf{u} \in U} \mathcal{H}\big(\boldsymbol{\xi}(t+1), \tilde{\mathbf{x}}(t), \mathbf{u}, t\big), \tag{3.75}$$

---

[1] The Greek letter λ is more commonly used than ξ; but we have already used λ to represent eigenvalues.

and at the final time T,

$$\xi'(T) = \eta' \frac{\partial \varsigma}{\partial \mathbf{x}} (\tilde{\mathbf{x}}(T)) + \frac{\partial L}{\partial \mathbf{x}} (\tilde{\mathbf{x}}(T)).$$

Note that if $\tilde{\mathbf{u}}$ is in the interior of II then condition (3.75) is equivalent to

$$\xi'(t+1) \delta \frac{\partial \mathbf{f}}{\partial \mathbf{u}} (\tilde{\mathbf{x}}(t), \tilde{\mathbf{u}}(t)) + \frac{\partial R}{\partial \mathbf{u}} (\tilde{\mathbf{x}}(t), \tilde{\mathbf{u}}(t)) = 0. \qquad (3.76)$$

If we apply this set of conditions to system (3.48) and define the matrices $\Phi(\mathbf{y})$ and $K(\mathbf{y})$ in terms of the derivatives of nonlinear functions, appearing in (3.48), as was done in the linearization Section 3.3.3 in equations (3.58) and (3.59), respectively, (3.74) can be written as

$$\xi'(t) = \xi'(t+1) \delta \left(G(\tilde{\mathbf{y}}(t)) + \Phi(\tilde{\mathbf{y}}(t)) + K(\tilde{\mathbf{y}}(t)) \right)$$
$$+ \frac{\partial R}{\partial \mathbf{x}} (\tilde{\mathbf{x}}(t), \tilde{\mathbf{u}}(t)), \qquad (3.77)$$

and equation (3.76) becomes

$$\xi'(t+1) = \frac{1}{\delta} \frac{\partial R}{\partial \mathbf{u}} (\tilde{\mathbf{x}}(t), \tilde{\mathbf{u}}(t)), \qquad (3.78)$$

where obviously $\tilde{\mathbf{y}}(t) = A\tilde{\mathbf{x}}(t)$. Note that equation (3.78) can be used to eliminate $\xi'(t+1)$ from (3.77).

3.4.4 Planning Horizons

A difficulty in formulating optimal yield problems is deciding on the appropriate specification of the final time condition (3.71). One approach is to specify that the resource should be at its maximum yield potential at the end of the planning interval $[0, T]$; that is,

$$\varsigma_i(\mathbf{x}(T)) = x_i(T) - x_i^o, \qquad i = 1, \ldots, k,$$

where x_i^o is the ith element of the MSY population vector $\mathbf{x}^o$. On the other hand, however, the problem can be avoided by allowing $T \to \infty$, in which case it is no longer necessary to specify $L(\mathbf{x}(T), T)$ in (3.68) or the function ς as discussed above. This is possible only in problems where the interest rate is greater than zero (that is, the discount factor $\delta < 1$). If $\delta = 1$, the sum of all discounted future rents, which is also referred to as the *present value* of the resource, has infinite value; that is,

$$J(\mathbf{x}_0) = \sum_{t=0}^{\infty} \delta^t R(\mathbf{x}(t), \mathbf{u}(t)) \tag{3.79}$$

is defined for $\delta = 1$ only if the rents themselves go to zero over time.

The infinite time formulation presents its own problems, however, since only analytical solutions to simple formulations or approximate numerically solutions to more complicated formulations are possible. Here we will briefly discuss how the finite and infinite time formulations, respectively involving (3.68) and (3.79), are related through the form of the function $L(\mathbf{x}(T), T)$ in (3.68). Although this discussion pertains to problems that are more general than those that are formulated just using nonlinear stage-structured models, we will subsequently apply the results obtained here to resources modeled by system (3.48).

The principle of optimality can be used to split the infinite time horizon problem into a finite time horizon problem that includes a transition period with length T to some terminal state $\mathbf{x}(T)$, and a problem that determines the optimal value for $\mathbf{x}(T)$. Specifically, if we let $\tilde{J}(\mathbf{x}_0)$ denote the solution to the problem

$$\max_{\substack{\mathbf{u}(t) \in U \\ t=0,1,2,\dots}} J(\mathbf{x}_0) = \sum_{t=0}^{\infty} \delta^t R(\mathbf{x}(t), \mathbf{u}(t))$$

subject to the equation (3.69) and initial condition (3.70), then it follows from the principle of optimality that

$$\tilde{J}(x_0) = \max_{\substack{u(t)\in U \\ t=0,1,2,\ldots}} \left(\sum_{t=0}^{T-1} \delta^t R(x(t), u(t)) \delta^T \sum_{t=T}^{\infty} \delta^{t-T} R(x(t), u(t)) \right)$$

$$= \max_{\substack{u(t)\in U \\ t=0,\ldots,T-1}} \left(\sum_{t=0}^{T} \delta^t R(x(t), u(t)) + \delta^T \tilde{J}(x_T) \right). \qquad (3.80)$$

For $\delta < 1$, the latter term in equation (3.80) is insignificantly small and can be ignored if T is sufficiently large. For relatively small T, rather than neglect this term altogether, we can replace it with its best approximation, assuming the system is in equilibrium from time T onwards. Thus if we force $x(t) = x_T$ for $t = T, T+1, \ldots$ and choose u_T to be the corresponding equilibrium value for u, that is, from equation (3.69)

$$x_T = f(x_T, u_T), \qquad (3.81)$$

then the term $\delta^T \tilde{J}(x_T)$ can be approximated by

$$L(x_T, T) = \frac{\delta^T}{1 - \delta} R(x_T, u_T).$$

In the previous subsection, we presented conditions under which the maximum yield solution to the dynamic nonlinear stage-structured problem incorporated the MSY solution in one time step. In this problem the control u appears linearly in both the system equation (3.48) and the value function (3.62) and thus linearly in the Hamiltonian function (3.73). If u appears nonlinearly in the Hamiltonian, then it is known that optimal solutions may asymptotically approach an equilibrium solution pair that satisfies equations (3.74) and (3.76) of the maximum principle for some constant vector ξ (Brock and

Scheinkman, 1976; Haurie, 1982); that is, $\mathbf{x}$, $\mathbf{u}$, and ξ must simultaneously satisfy

$$\mathbf{x} = \mathbf{f}(\mathbf{x}, \mathbf{u}),$$

$$\xi'\left(\delta\frac{\partial\mathbf{f}}{\partial\mathbf{x}}(\mathbf{x}, \mathbf{u}) - I\right) + \frac{\partial R}{\partial\mathbf{x}}(\mathbf{x}, \mathbf{u}) = 0, \qquad (3.82)$$

and

$$\xi'\delta\frac{\partial\mathbf{f}}{\partial\mathbf{u}}(\mathbf{x}, \mathbf{u}) + \frac{\partial R}{\partial\mathbf{u}}(\mathbf{x}, \mathbf{u}) = 0.$$

An equivalent set of equations has been derived by Knapp (1983) using a dynamic programming formulation.

From elementary calculus it is known that the second and third equations in (3.82) provide a set of necessary conditions for $(\mathbf{x}, \mathbf{u})$ to maximize $R(\mathbf{x}, \mathbf{u})$ subject to the constraint $\mathbf{x} = \delta f(\mathbf{x}, \mathbf{u})$; that is, as $\delta \to 1$ the solution $(\mathbf{x}, \mathbf{u})$ to (3.82) approaches a solution to the maximum sustainable rent problem

$$\max_{\mathbf{u}} R(\mathbf{x}, \mathbf{u}) \quad \text{subject to} \quad \mathbf{x} = \mathbf{f}(\mathbf{x}, \mathbf{u}). \qquad (3.83)$$

From the above discussion it is apparent that two methods can be used to approximate $\tilde{J}(\mathbf{x}_0)$ expressed in (3.80). The first is to solve

$$\max_{\substack{\mathbf{u}(t)\in U \\ t=0,\dots,T}} \left(\sum_{t=0}^{T-1} \delta^t R(\mathbf{x}(t), \mathbf{u}(t)) + \frac{\delta^T}{1-\delta}R(\mathbf{x}(T), \mathbf{u}_T)\right) \qquad (3.84)$$

subject to equations (3.69), (3.81), and $\mathbf{u}_T = \mathbf{u}(T)$. The second is to solve

$$\max_{\substack{\mathbf{u}(t)\in U \\ t=0,\dots,T-1}} \left(\sum_{t=0}^{T-1} \delta^t R(\mathbf{x}(t), \mathbf{u}(t)) + \frac{\delta^T}{1-\delta}R(\mathbf{x}_T, \mathbf{u}_T)\right)$$

subject to equation (3.69) and the pair $(\mathbf{x}_T, \mathbf{u}_T)$ must satisfy equations (3.82); note that the first formulation involves optimizing over $\mathbf{u}(T)$ while the second does not.

Since the second problem is a constrained version of the first (it has an extra constraint that comes about when eliminating ξ from the second and third equation in (3.82)), its solution will be an inferior approximation to an infinite time problem. Also note that the second problem may not have a solution if no controls $u \in U$ exist that will drive the system to the constraint endpoint $x(T)$ over the allocated time interval T.

Finally, note that for the nonlinear stage-structured system modeled by equation (3.48), the costate variable ξ is easily eliminated from the last two equations in (3.82) (cf. equations (3.77) and (3.78)) to obtain

$$\frac{1}{\delta}\frac{\partial R}{\partial u}(x, u) = \left(I - \delta\left(G(y) + \Phi(y) + K(y)\right)\right)^{-1}\frac{\partial R}{\partial x}(x, u).$$
(3.85)

Note that equation (3.85) is a condition that is satisfied only by interior optimal solutions, that is, for $u(t)$ not on the boundary of U.

3.5   STOCHASTIC THEORY

3.5.1   Introduction

In the previous chapter we remarked that in the interests of realism, we need to consider nonlinear stochastic models. We also noted that stochastic theories are rather complex, and comprehensive treatments are beyond the scope of this book. In this vein, we will begin fairly modestly and quit when the going gets too rough. Along the way, we will obtain some useful models, insights, and results.

3.5.2   Stochastic Recruitment—Small Noise

In Section 3.2 we emphasized that density-dependent mechanisms are perhaps most important in determining survival among newborn individuals, because the youngest

103

individuals are generally the most vulnerable cohort of individuals. For the same reasons, environmental stochasticity and variation in the ability of individuals to survive is most strongly expressed in the survival of individuals in this age group. Thus we will begin our stochastic analysis by assuming that the only stochastic parameter in a model such as (3.44) is the initial survival parameter s_0. Under this assumption, recruitment at time t is a stochastic variable $X_1(t)$, and all other stage-class variables $X_i(t)$, as well as the aggregated newborn or egg variable $X_0(t)$, are stochastic by virtue of the fact that they are proportions (see equation (3.7)) or linear combinations (equation (2.10)) of stochastic variables.

For the sake of simplicity, consider the age-structured model (3.26), rather than the stage-structured model (3.44), and assume that the only stochastic parameter in (3.26) is s_0 (analysis of the more complicated case follows analogously). The most convenient way to express the stochasticity of s_0 is to multiply it by random variables Z_t, where, for all t, these variables are independent, and are drawn from the identical distribution (that is, $i.i.d.$ variables) for which (recall (2.51) for a definition of the variance operator $\mathcal{V}$)

$$E[Z_t] = 1 \quad \text{and} \quad \mathcal{V}[Z_t] = \sigma_Z^2. \qquad (3.86)$$

Thus consider the model

$$X_0(t) = \sum_{i=1}^{n} b_i X_i(t)$$
$$X_1(t+1) = Z_t \phi\big(X_0(t)\big) \qquad (3.87)$$
$$X_{i+1}(t+1) = s_i X_i(t), \qquad i = 1, \ldots, n-1,$$

where ϕ is the total recruitment function (linear recruitment function multiplied by the nonlinear modifier function ψ).

To obtain a set of equations for the vector of means $\bar{x}(t)$, we can calculate the expectations associated with equations (3.87). The only difficulty associated with this operation is calculating $E[X_1(t+1)]$, since this involves the expectation of the product of a random variable and a nonlinear function of a random variable. In general, to carry out the expectation operation, we need to make some type of approximation. If we assume that σ_Z^2 is small, so that third and higher order moments around the mean can be neglected, then it follows from calculus (Taylor expansions) and elementary probability theory that (using the notation γ_{ii} of the previous chapter to denote the variance associated with the stochastic variable X_i)

$$E\left[\phi(X_0)\right] \approx \phi(\bar{x}_0) + \frac{1}{2}\gamma_{00}\frac{d^2\phi}{dx^2}(\bar{x}_0). \qquad (3.88)$$

Under the same *small noise* assumption, the variance approximation

$$\mathcal{V}[\phi(X_0)] \approx \left(\frac{d\phi}{dx}(\bar{x}_0)\right)^2 \gamma_{00}$$

will be used to calculate the variances associated with system (3.87). Before being able to find the means and variance/covariance terms associated with this system we need the following two results. (For a derivation see Rao, 1973. Also note that the subscript to E indicates the variable over which the expectation is taken, and the variable on which the expectation is conditioned is indicated by the subscript to the vertical line $|$.)

$$E[X_1(t+1)] = E_Z\left[Z_t E_{X_0}\left[\phi(X_0(t))|z\right]\right]$$
$$= E_{X_0}\left[\phi(X_0(t))\right],$$

since Z_t is independent of $X_0(t)$ and $E[X_0] = 1$; and (see Getz, 1984a)

$$\mathcal{V}\left[Z_t\phi(X_0(t))\right] = \mathcal{V}_Z\left[E_{X_0|z}\left[Z_t\phi(X_0(t))\right]\right] \qquad (3.89)$$

105

$$+ E_Z\left[V_{X_0|z}\left[Z_t\phi(X_0(t))\right]\right]$$
$$= E\left[\phi(\bar{x}_0(t))\right]\sigma_Z^2 + V\left[\phi(\bar{x}_0(t))\right](\sigma_z^2 + 1).$$

If we now use the expectation operator to take the first and second order moments associated with system (3.87), then (recalling definition (2.50)) it follows from equations (3.88) and (3.89) that the means and variances of the process are updated by

$$\bar{x}_1(t+1) \approx \phi(\bar{x}_0(t)) + \frac{1}{2}\frac{d^2\phi}{dx^2}(\bar{x}_0(t))\gamma_{00}(t),$$
$$\bar{x}_{i+1}(t+1) = s_i\bar{x}_i(t), \quad i = 1,\ldots,n-1, \qquad (3.90)$$
$$\gamma_{11}(t+1) \approx \phi(\bar{x}_0(t))^2\sigma_Z^2 + \left(\frac{1}{2}\frac{d\phi}{dx}(\bar{x}_0(t))\right)^2\gamma_{00}(t),$$

$$\gamma_{1\,j+1}(t+1) \approx s_j\gamma_{0j}(t)\frac{d\phi}{dx}(\bar{x}_0(t)), \quad j = 1,\ldots,n-1,$$
$$\gamma_{i+1\,j+1}(t+1) = s_i s_j \gamma_{ij}, \qquad i,j = 1,\ldots,n-1,$$

where

$$\bar{x}_0(t) = \sum_{i=1}^{n} b_i \bar{x}_i(t),$$

$$\gamma_{00}(t) = \sum_{i=1}^{n}\sum_{j=1}^{n} b_i b_j \gamma_{ij}(t),$$

$$\gamma_{0j} = \sum_{i=1}^{n} b_i \gamma_{ij}, \qquad j = 1,\ldots,n-1.$$

More generally, if we carry out the same set of calculations for the stage-structured model (3.44), we obtain a system of equations that can be compactly expressed using the Kronecker matrix product defined in (2.52). Specifically, defining the $n+1$-dimensional vector, $\boldsymbol{\gamma}_0 = (\gamma_{00},\ldots,\gamma_{0n})$, the stochastic equivalent of system (3.44) is

$$\begin{pmatrix} \bar{\mathbf{x}}(t+1) \\ \boldsymbol{\gamma}(t+1) \end{pmatrix} = \begin{pmatrix} G & \mathbf{0} \\ \mathbf{0} & G\otimes G \end{pmatrix}\begin{pmatrix} \bar{\mathbf{x}}(t) \\ \boldsymbol{\gamma}(t) \end{pmatrix} + \varphi(\bar{x}_0(t), \boldsymbol{\gamma}_0(t)), \quad (3.91)$$

where all the elements of the $n(n+1)$-dimensional vector are zero except for the $n+1$ elements

$$\varphi_1 = \phi(\bar{x}_0) + \frac{1}{2}\frac{d^2\phi}{dx^2}(\bar{x}_0)\gamma_{00}$$

$$\varphi_{n+1} = \phi(\bar{x}_0)^2\sigma_Z^2 + \left(\frac{1}{2}\frac{d\phi}{dx}(\bar{x}_0)\right)^2\gamma_{00}$$

$$\varphi_{n+1+j} = s_j\gamma_{0j}\frac{d\phi}{dx}(\bar{x}_0(t)), \quad j = 1,\ldots,n-1.$$

If we compare system (3.91) to system (2.60), we notice that all entries in the lower left submatrix of (3.91) are zero. This is a consequence of the fact that all the elements of the matrix G are considered to be constant, whereas in system (2.60) the elements of the matrix A are either probabilities or stochastic variables.

The techniques used to derive system (3.91) can also be used to derive a set of stochastic equivalents for the full aggregated nonlinear stage-structured model (3.48). In this case, however, because the elements of the system matrix G are now functions of the stochastic aggregated variables $Y = \Theta X$, the function approximation technique will have to be applied to each element of G, thereby leading to a much more complicated set of dynamic equations.

Note that, by removing the argument t in system (3.91), the means $\bar{x}$ and variance/covariance terms γ of the steady state distribution of the random variables X can be solved for (also see Reed, 1983). This distribution will be approached with time only if system (3.91) is stable. Pollard (1973), however, proves that if λ_i, $i = 1,\ldots,n$ are the eigenvalues of a (general) matrix G, then the n^2 eigenvalues of the matrix $G \otimes G$ have the values $\lambda_i\lambda_j$, $i,j = 1,\ldots,n$. Therefore, if the deterministic system associated with G is stable (all eigenvalues lie within the unit circle), then the extended stochastic system (3.91) is stable.

3.5.3 Stochastic Recruitment—Large Noise

We now return to the system equations (3.87) satisfied by the stochastic vector X. If σ_Z^2 (see (3.86)) is around $\frac{1}{2}$ or larger, as is the case in most marine fisheries, then approximations (3.88) and (3.12) can be highly inaccurate and a different approach is required.

One approach is to assume all stochastic variables X_j satisfy a particular distribution density function $\mathcal{F}(\bar{x}_j, \gamma_{jj}; \varepsilon_j)$ that is fully determined once $\bar{x}_j$ and γ_{jj} are known ($j = 0, \ldots, n,$). Since the variables X_j are all distributed on $[0, \infty)$, $j = 0, \ldots, n$ it follows that

$$Prob\{X_j \leq x_j\} = \int_0^{x_j} \mathcal{F}(\bar{x}_j, \gamma_{jj}; \varepsilon_j) d\varepsilon_j. \qquad (3.92)$$

The most obvious distributions to use are the lognormal or gamma, since both are defined on $[0, \infty)$ and are described by two parameters.

Suppose from a set of data that we are able to estimate a mean stock-recruitment function $\bar{\phi}(x_0)$ and an associated coefficient of variation $\mu(x_0)$; that is,

$$\bar{\phi}(x_0) = E[X_1(t+1)|X_0 = x_0],$$

and

$$\mu(x_0) = \mathcal{V}[X_1(t+1)|X_0 = x_0]/\bar{\phi}(x_0)^2.$$

Before we write down new expressions for the approximate equations in (3.90) we need to define the statistics $x_{0|j}$ and $\gamma_{00|j}$ respectively as the mean and variance of X_0 given $X_j = x_j$ for one $j \in \{1, \ldots, n\}$, but the remaining X_i are random (only their means and variances are known); that is,

$$x_{0|j} = \sum_{\substack{i=1 \\ i \neq j}}^{n} b_i \bar{x}_j + b_j x_j,$$

$$\gamma_{00|j} = \sum_{\substack{i=1 \\ i \neq j}}^{n} \sum_{\substack{k=1 \\ k \neq j}}^{n} b_i b_k \gamma ik.$$

There are a number of technical details associated with taking the expectations to derive new approximate expressions for $\bar{x}_1(t+1)$, and $\gamma_{1j}(t+1)$, $j = 1, \ldots, n$, in equations (3.90). For details the reader is referred to Getz (1984a). The expressions are:

$$\bar{x}_1(t+1) \approx \int_0^{x_j} \phi(\varepsilon_0) \mathcal{F}(\bar{x}_0(t), \gamma_{00}(t); \varepsilon_0) d\varepsilon_0,$$

$$\gamma_{11}(t+1) \approx \int_0^{\infty} \mu(\varepsilon_j)(1 + \bar{\phi}(\varepsilon_0)^2) \mathcal{F}(\bar{x}_0(t), \gamma_{00}(t); \varepsilon_0) d\varepsilon_0$$
$$- \bar{x}_1(t+1)^2,$$

$$\gamma_{1\,j+1}(t+1) \approx s_j \int_0^{\infty} \varepsilon_j \mathcal{F}(\bar{x}_j(t), \gamma_{jj}(t); \varepsilon_j)$$
$$\times \int_0^{x_j} \bar{\phi}(\varepsilon_{0|j}) \mathcal{F}(\bar{x}_{0|j}(t), \gamma_{00|j}(t); \varepsilon_{0|j}) d\varepsilon_{0|j} d\varepsilon_j,$$
$$j = 1, \ldots, n - 1. \tag{3.93}$$

Consider the case where the probability density functions (3.92) are assumed to be gamma; that is (dropping the subscript j),

$$\mathcal{F}(\bar{x}, \gamma; \varepsilon) = \frac{\bar{x}/\gamma}{\Gamma(\bar{x}^2/\gamma)} \left(\frac{\bar{x}}{\gamma} \varepsilon\right)^{\frac{\bar{x}^2}{\gamma} - 1} e^{-\frac{\bar{x}}{\gamma}\varepsilon}, \tag{3.94}$$

where $\Gamma(\cdot)$ is the gamma function (extension of the factorial function to noninteger values). If the recruitment function is either Ricker (see expression (3.10)), or has the form

$$x_1 = \frac{s_0}{\beta}\left(1 - e^{-\beta x_0}\right)$$

(this is the same basic compensatory shape as the Beverton and Holt function (3.9)), then analytical expressions exist for the first two integrals in (3.93). For details see Getz (1984a).

3.5.4 Stochastic Harvesting

The appropriate way to incorporate harvesting into a stochastic model depends very much on the problem at hand. One can assume either that a known number of individuals are removed from each stage class in each time period, in which case the yield is deterministic, or that this number is itself a stochastic variable.

A common assumption made in fisheries management is that the harvest from a particular age class is proportional to a fishing intensity variable and the number of individuals in that age class. This approach will be analyzed in Chapter 4. Here we will confine ourselves just to making some general comments relating to the management of a resource system modeled by stochastic nonlinear models.

In general, stochastic optimization problems associated with maximizing rent subject to system equations are much harder to analyze than their deterministic counterparts. Because equation (3.48) is easily linearized to obtain equation (3.57), however, a number of techniques, such as Kalman filtering, are available to compute approximate optimal solutions. The reader is referred elsewhere for a treatment of the theory of stochastic control systems (Goodwin and Sin, 1984).

A very important notion associated with the management of stochastic systems is that of feedback or adaptive management. In deterministic systems, one can completely specify the management policy in terms of the values of the controls $\mathbf{u}(t)$ to be applied during each future time period, because the system will behave exactly as predicted by the model. In stochastic systems, however, the system will deviate from the predicted values, and this deviation will increase with time. Thus it is important to "update" the management strategy at the beginning of each time period according to the best estimate of what the current value of the state is. If it is possible to characterize the optimal management strategy purely in terms of the state of the system, then the optimal value for $\mathbf{u}(t)$ can be directly calculated from the estimate of the state. Any control that depends on the value of the state is termed a feedback or adaptive control.

Adaptive management of resource systems has been studied quite intensively over the past 10 to 15 years (for a comprehensive review, see Walters, 1986). One can distinguish between passive and active adaptive prescriptions for management strategies, where active prescriptions include evaluating the amount of information that is obtained from implementing a particular management strategy with respect to improving estimates of parameters in the resource model. In theory, active adaptive management has great merit. In practice, its policies may lead to strategies that increase yield variability between years (current yield may be sacrificed to increase estimated future yields) and this may be unacceptable when considering some of the broader socioeconomic issues related to the problem, as discussed in the next chapter.

3.5.5   *Level of Complexity*

In Chapter 2 we focused on linear models for which the parameters of the system transition matrix A (see equation (2.59)) were regarded as representing the probability that certain events occur. In this chapter we have thus far dealt with the more general problem of stochastic model parameters, but in the restrictive context of the recruitment process. Obviously these ideas can be extended to consider stochastic survival, as will be done in Chapter 4 in the context of fisheries problems. Stochasticity and uncertainty enter resource management systems in several additional ways, however, that make it extremely difficult to analyze management problems associated with these systems. As a result, most of the analysis pertaining to stochastic resource management problems is done in the context of scalar models that do not include age or class structures in the resource (for example, see Walters, 1986). When age structure is added, only narrowly focused questions are dealt with easily.

Uncertainty enters resource management problems in at least four fundamentally different ways. The first is variability in the environment. In every model we are required to define those variables that are an intrinsic part of the resource (system variables) and those variables that impact but are not impacted by the system variables (environmental variables). Such a distinction is invariably artificial, but the process of model building requires the identification of a limited number of system variables. The behavior of the environmental variables can be assumed to be determined, but most often these variables are best characterized as stochastic (e.g., temperature and rainfall).

A second source of uncertainty is that the models themselves (unlike many physical systems models) are poor approximations to reality. In many cases we cannot confidently predict the first significant digit of population density or yield. Further, we are often unable to perform experiments that allow us to determine the accuracy of our models.

The third source of uncertainty is that, in many resource systems, there are problems associated with measuring the state of the resource while the management process takes place. This is especially true for marine resources.

Finally, even the management aspect of the problem is an idealized formulation of the true management problem. In some cases management has to deal with conflicts that arise between parties that have different interests in the resource, and the management solution represents a subjective compromise on the part of the manager. Thus the question of what is the most appropriate management formulation provides a fourth area of uncertainty in our analysis.

In subsequent sections we will deal with some of these issues in the context of specific problems that relate to age-structured resources, although we will not attempt to address questions relating to broader socioeconomic issues. For detailed discussion of questions relating to uncertainty in resource systems, the reader is referred to Mangel (1985) and Walters (1986).

3.5.6 Monte Carlo Methods

We have seen how analytical techniques can be used to construct exact or approximate expressions for the evolution of the first and second order moments of the probability distributions associated with systems of equations such as (2.59) or (3.87). This approach is severely limited when it comes to complex nonlinear systems involving several stochastic parameters. A general means of investigating the behavior of complex stochastic systems is through the application of Monte Carlo simulation techniques. This technique takes its name from the premier city of cards, dice, and roulette wheels, where an evening's entertainment is an experiment in probability theory. The Monte Carlo method is equivalent to carrying out a numerical experiment consisting of a large number of simulation runs. In each run, stochastic parameters are assigned particular values that are randomly drawn afresh at each time step from appropriate probability distributions. Data obtained from these runs are used to estimate the moments (or percentiles) of probability distributions associated with the results of the simulations (for example, expected yield or probability that the stock index falls below a predetermined level).

To be more specific, consider system (3.87). For a given set of initial conditions $X(0)$, the time sequence of the stochastic vector $X(t)$, $t = 1, \ldots, T$ has a particular realization that is determined by the specific value of Z_t that arises at each point in time. If Z_t is determined by the throw of a dice or the spin of a wheel, then over a large number of runs the average outcome per run—for example, the final stock index realized by $X_0(T)$—is a function both of the form of the model and the type of dice used to determine Z_t. If we know that Z_t has an equal probability of taking on any value between 0 and 2, then a 21-faced dice could be used to assign a value of Z_t to 1 decimal place ($1 \equiv 0.0$, $2 \equiv 0.1, \ldots, 21 \equiv 2.0$) or, better still, a 201-faced dice could be used to assign a value to Z_t to 2 decimal places, for each t of a simulated run.

Dice simulation computer software algorithms are not truly random and thus the values generated are referred to as pseudo-random. There are many algorithms available for generating sets of integers lying between 0 and n where each integer value has an equal probability of occurring. These values can be scaled to obtain a number lying on the interval $[0, 1]$. For convenience $\mathcal{U}(0, 1)$ will be used to denote the uniform distribution on $[0, 1]$; that is, $U \in \mathcal{U}(0, 1)$ is a random variable U equally likely to take on any particular value in the interval $[0, 1]$. Almost all computers have software available for simulating the throwing of an n-faced dice and generating from these throws the distribution $\mathcal{U}(0, 1)$. Many computer systems also have the software required to generate the normal, lognormal, and numerous other distributions, and some of these distributions can be generated in several different ways. Here we present a small selection of algorithms that can be used to address problems relating to the management of stochastic fisheries, as discussed in Chapter 4. Readers are cautioned, however, to check that the random numbers obtained from their particular software do not exhibit cycles within the length of the sequence of pseudo-random num-

114

bers being generated, and to ensure that different runs are made with distinct sequences of pseudo-random numbers.[2]

Perhaps the most commonly used pseudo-random number generator is a congruential method based on the modulus function to integer values. Specifically, any integer I can be expressed as $I = nm + r$, where $0 \leq r \leq m - 1$ is the integer remainder r and n is the maximum possible integer number of times that I can be divided by m and still leave a positive remainder. The value of $I \pmod{m}$ is the remainder r. The congruential method then is used to generate sequences of a pseudo-random variable X_t using the relationship

$$X_{t+1} = (aX_t + c) \pmod{m}, \qquad (3.95)$$

where the *multiplier* a, the *increment* c, and the *modulus* m are nonnegative integers. For a "randomly" selected starting or *seed* value X_0, (3.95) yields a sequence of values $\{X_t\}$ that will eventually begin to repeat itself after $t = p \leq m$ iterations. Random numbers $U_t \in \mathcal{U}(0, 1)$ are generated using the transformation $U_t = X_t/m$. It has been noted (see Rubenstein, 1981) that a good set of constants for a binary computer are $m = 2^{35}$, $a = 2^7 - 1$, and $c = 1$.

A number of algorithms exist for generating random variables of various distributions (normal, lognormal, exponential, gamma, etc.), using variates $U_1, U_2, \ldots$, drawn from $\mathcal{U}(0, 1)$. Here we briefly summarize three of those that have application to population modeling problems. For a complete treatment of these algorithms the reader is referred to Rubenstein (1981), as well as Swartzman and Kaluzny (1987) for examples of population modeling applications.

---

[2] For a more general discussion the reader is referred to Rubenstein (1981) or one of the many texts available on this topic.

Standard Normal Variates. A relatively simple and fast technique for generating variates $Z_t \in N(0, 1)$, the standard normal distribution, is due to Box and Muller (see Rubenstein, 1981).

Algorithm

1. Generate variates U_1 and U_2 belonging to $\mathcal{U}(0, 1)$.
2. Calculate

$$Z_1 = (-2 \ln U_1)^{1/2} \cos 2\pi U_2,$$

$$Z_2 = (-2 \ln U_1)^{1/2} \sin 2\pi U_2.$$

3. Output Z_1 and Z_2 as normal variates satisfying $E[Z] = 0$ and $\mathcal{V}[Z] = 1$.

This algorithm follows from Box and Muller (see Rubenstein, 1981) who proved that if U_1 and U_2 are independent random variates from $\mathcal{U}(0, 1)$, then Z_1 and Z_2 are independent random variates from $N(0, 1)$.

Once variates from $N(0, 1)$ have been generated, then a simple transformation can be used (see step 2 in the next algorithm) to generate variates belonging to $N(\mu, \sigma^2)$, the normal distribution with mean μ and variance σ^2.

Lognormal Variates. The lognormal distribution has probability density function

$$\mathcal{F}(y) = \frac{1}{\sqrt{2\pi\sigma^2}} \frac{1}{y} e^{-\frac{(\ln y - \mu)^2}{2\sigma^2}}, \qquad 0 \leq y < \infty. \qquad (3.96)$$

where $\mathcal{F}(y)$ is defined to be zero whenever $y < 0$. Variates Y satisfying the lognormal distribution (3.96) are easily generated as follows (Rubenstein, 1981).

Algorithm

1. Generate variate Z belonging to $N(0, 1)$.
2. Calculate variate $X = \mu + \sigma Z$ (which belongs to $N(\mu, \sigma^2)$).
3. Calculated variate $Y = e^X$ (which is now distributed according to $\mathcal{F}(y)$ given in (3.96)), where the mean of these variates is

$$E[Y] = e^{\mu + \frac{1}{2}\sigma^2}$$

and the variance is

$$\mathcal{V}[Y] = e^{2\mu + \sigma^2}(e^{\sigma^2} - 1).$$

Note that the median of the probability distribution $\mathcal{F}(y)$ given in (3.96) is e^μ, as might be expected from the transformation $Y = e^X$. The mean, however, is shifted to the right of the median by the factor $e^{\frac{1}{2}\sigma^2}$ as a consequence of the asymmetry of the log transformation in condensing $(-\infty, 0]$ onto $[0, 1]$.

Beta Variates. As discussed in Section 2.4.3, the parameters associated with the system matrix A may themselves be stochastic variables. Since survival and transition parameters take on values between 0 and 1, it is also useful to be able to generate stochastic variables defined on $[0, 1]$ that have a specific mean and variance. The beta distribution, $\mathcal{B}_{\alpha,\beta}(y)$, is defined for $y \in [0, 1]$ by

$$\mathcal{B}_{\alpha,\beta}(y) = cy^\alpha(1 - y)^\beta \qquad 0 \le y \le 1, \qquad (3.97)$$

where the constant

$$c = \frac{\Gamma(\alpha)\Gamma(\beta)}{\Gamma(\alpha + \beta)},$$

117

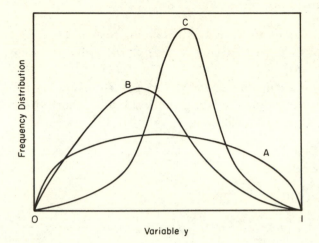

FIGURE 3.2. Typical shape of the beta distribution (see equation (3.97)) for the cases where A—$\alpha < 1$, $\beta < 1$; B—$\alpha = 1$, $\beta > 1$; C—$\alpha > 1$, $\beta > 1$.

(defined in terms of the gamma function $\Gamma(\cdot)$; also see equation (3.94)) ensures that

$$\int_0^1 B_{\alpha,\beta}(y)dy = 1.$$

The shape of this distribution, as illustrated in Figure 3.2, depends on the values of the parameters α and β.

The following procedure generates a stochastic variable Y drawn from the two parameter beta distribution $B(\alpha, \beta)$.

Algorithm

1. Generate variates U_1 and U_2 belonging to $\mathcal{U}(0, 1)$.
2. Calculate $X_1 = U_1^{1/\alpha}$.
3. Calculate $X_2 = U_2^{1/\beta}$.
4. If $X_1 + X_2 \geq 1$ discard values and begin again; otherwise calculate variate $Y = \frac{X_1}{X_1 + X_2}$.

118

5. Output Y as a variate with mean

$$E[Y] = \frac{\alpha}{\alpha + \beta},$$

(which belongs to $[0, 1]$ as long as α and β are both positive), and variance

$$\mathcal{V}[Y] = \frac{\alpha\beta}{(\alpha + \beta)^2(\alpha + \beta + 1)}.$$

Examples of how the above procedures are used in Monte Carlo simulations are presented in Chapter 4.

3.6 AGGREGATED AGE STRUCTURE

3.6.1 Model

Perhaps the most challenging problem in applying age- or, more generally, stage-structured models to resource problems is obtaining reasonable estimates for the model parameters. In forest-stand management problems, mortality rates and growth (hence stage transition) rates can be observed directly from time sequences of measurements taken with respect to particular stands. On the other hand, fish populations are difficult to observe directly, and mortality rates and recruitment levels are often inferred using retrospective procedures known as Virtual Population Analysis (VPA; see Pope, 1972). These procedures use catch-at-age (how many fish of a particular age are in the catch) and fishing effort data, but their reliability is questionable (Fournier and Archibald, 1982; Bergh, 1986). In many situations, there is insufficient data or the variability in the data is too large to robustly estimate more than three or four parameters from a given set of data. In this case, the most appropriate course of action is to assume that mortality is characterized by a single age-independent survival constant $s_i = s$, $i = 0, \ldots, n$.

(Note that if the first year-survival parameter s_0 is actually much less than s, as it often is, then the ensuing analysis still holds if we multiply the recruitment modifier function $\psi(x_0)$ by s_0/s; see equation (3.6)). With this restriction, the elements l_i defined by expression (2.2) are simply

$$l_i = s^i, \qquad i = 0, 1, \ldots. \qquad (3.98)$$

For generality, we will consider the case where the stock-recruitment relationship is delayed by r years; that is, we will assume that relationship (3.15) holds. Note that recruitment at age r implies that r is the first age at which individuals can be observed (usually in the catch) so that data are available to construct a relationship between recruits and the fecund stock r years ago.

As in deriving relationship (3.7), it follows that

$$x_i(t + 1) = \frac{l_i}{l_r} x_r \big(t - (i - r)\big). \qquad (3.99)$$

Hence, the following delay-difference equation can be derived from (3.15) and assumption (3.98):

$$x_0(t + 1) = \sum_{i=1}^{\infty} b_i s^{i-r} \phi(x_0(t - i)). \qquad (3.100)$$

We will examine two cases associated with assumptions that simplify the set of natality parameters b_i. The first applies to organisms, such as many mammals, where all reproductively mature females give birth to the same expected number of progeny, irrespective of their particular age; that is, for some $b > 0$,

Case 1.

$$b_i = \begin{cases} 0, & i = 1, \ldots, i_0 - 1; \\ b, & i = i_0, i_0 + 1, \ldots \end{cases}.$$

Note we assume that $i_0 \geq r$, since only age classes i_0 onwards appear in the analysis. Thus Case 1 implies that no individuals reproduce until age i_0, at which time they all reproduce at the same rate.

In contrast, the number of eggs spawned by the females of many fish species is often correlated with the size of the female, and it is realistic to at least link natality parameters, say, to weight. In this case we assume that

Case 2.

$$b_i = \begin{cases} 0, & i = 1, \ldots, i_0 - 1; \\ bw_i, & i = i_0, i_0 + 1, \ldots, \end{cases} \tag{3.101}$$

where b is a natality rate per unit biomass and w_i is the biomass of a female aged i. A set of w_i may be obtained for a given population. However, here we will consider populations for which w_i satisfy a growth equation

$$w_{i+1} = c_1 w_i + c_2 w_{i-1}, \qquad i = 2, 3, \ldots. \tag{3.102}$$

A special case of this equation is Ford's growth equation (see Schnute and Fournier, 1980),

$$w_{i+1} - w_i = c(w_i - w_{i-1}), \qquad i = 2, 3, \ldots, \tag{3.103}$$

where the growth parameter c typically satisfies $0 < c < 1$. (Note that equation (3.104) is obtained from equation (3.102) by setting $c_1 = 1 + c$ and $c_2 = -c$.) Ford's equation has the interpretation that, from one year to the next, the net increase in the weight of each individual decreases by a constant factor $c \leq 1$. Rather than regarding w_0 in equation (3.103) as the weight of a "zero"-year-old, it may be more appropriate to regard both w_0 and c as parameters that can be selected to fit a set of empirical weight-at-age values $\{w_1, w_2, \ldots, w_n\}$.

The analysis that now follows is carried out for Case 2 above. Note that Case 1 is just a special case of Case 2 with $c = 1$ in Ford's equation (3.103), and the units of b appropriately adjusted. The same methodology employed could be applied to problems where the weight parameters satisfy a higher order difference equation than (3.102) or fecundity is independent of weight but the survival rates s_i are expressed as a difference relationship rather than being constant.

From assumption (3.101), equation (3.100) can be written as

$$x_0(t+1) = b \sum_{i=i_0}^{\infty} w_i s^{i-r} \phi\big(x_0(t+1-i)\big). \qquad (3.104)$$

If we now take the first term ($i = i_0$) out of the sum, and renumber the indices by adding 1 to each index, this equation becomes

$$
\begin{aligned}
x_0(t+1) &= b w_{i_0} s^{i_0-r} \phi\big(x_0(t+1-i_0)\big) \\
&\quad + b \sum_{i=i_0}^{\infty} w_{i+1} s^{i-r+1} \phi\big(x_0(t-i)\big) \\
&= b w_{i_0} s^{i_0-r} \phi\big(x_0(t+1-i_0)\big) \\
&\quad + c_1 b s \sum_{i=i_0}^{\infty} w_i s^{i-r} \phi\big(x_0(t-i)\big) \\
&\quad + c_2 b s^2 \sum_{i=i_0}^{\infty} w_{i-1} s^{i-1-r} \phi\Big(x_0\big(t+1-(i-1)\big)\Big),
\end{aligned}
$$

using equation (3.102);

$$
\begin{aligned}
&= c_1 s x_0(t) + c_2 s^2 x_0(t-1) \\
&\quad + b w_{i_0} s^{i_0-r} \phi\big(x_0(t+1-i_0)\big) \\
&\quad + b w_{i_0-1} s^{i_0-r+1} c_2 \phi\big(x_0(t-i_0)\big), \qquad (3.105)
\end{aligned}
$$

using equation (3.104) to replace summations.

In the case of Ford's equation (3.103) with $i_0 = r$ (see Schnute, 1985), equation (3.105) becomes

$$x_0(t + 1) = (1 + c)sx_0(t) - cs^2 x_0(t - 1)$$
$$+ bw_r \phi\big(x_0(t + 1 - r)\big)$$
$$- cbw_{r-1} s\phi\big(x_0(t - r)\big). \qquad (3.106)$$

Also, in this equation if we set $b = 1$, then x_0 is a stock biomass (cf. definition (3.101)) rather than the number of newborn individuals or an egg-production index.

Since the stock dynamics equation (3.106) has no age structure, harvesting cannot be age specific. Thus we assume that a constant proportion $h(t)$ of individuals is removed from each recruited age class (which comprises the aggregated stock variable $x_0(t)$). In keeping with the convention adopted in Figure 2.2, we assume that individuals are harvested after natural mortality has occurred but before reproduction takes place on the time interval $[t, t+1]$ (see Schnute, 1985, for a different timing convention). Under these assumptions it follows that equation (3.99) has the form

$$x_i(t + 1) = \left(\prod_{j=1}^{i-r+1} [1 - h(t + 1 - j)] \right) \frac{l_i}{l_r} x_r\big(t - (i - r)\big).$$
$$(3.107)$$

If this relationship is applied to the derivation of equation (3.105) we obtain

$$x_0(t + 1) = c_1[1 - h(t)]sx_0(t)$$
$$+ c_2[1 - h(t)][1 - h(t - 1)]s^2 x_0(t - 1)$$
$$+ bw_{i_0} \prod_{j=1}^{i_0-r+1}[1 - h(t + 1 - j)]s^{i_0 - r} \phi\big(x_0(t + 1 - i_0)\big)$$
$$+ c_2 bw_{i_0-1} \prod_{j=1}^{i_0-r+2}[1 - h(t + 1 - j)]s^{i_0 - r+1} \phi\big(x_0(t - i_0)\big).$$
$$(3.108)$$

123

The biomass yield obtained in each time interval is, of course,

$$J\big(h(t)\big) = h(t) \sum_{i=r}^{\infty} w_i s x_i(t),$$ (3.109)

since harvesting a proportion $h(t)$ of the population takes place after natural mortality has reduced the number of individuals in the ith age class from $x_i(t)$ to $sx_i(t)$. Splitting the sum in expression (3.109) at $i_0 \geq r$, it follows from stock-recruitment relationship $x_r(t+r) = \phi(x_0(t))$ (that is, relationship (3.15)), and relationships (3.98), (3.104), and (3.107) that

$$J\big(h(t)\big) = h(t)\Bigg(\frac{sx_0(t)}{b}$$
$$+ \sum_{i=r}^{i_0-1} w_i s^{i-r+1} \prod_{j=1}^{i-r+1} [1 - h(t-j)]\phi\big(x_0(t-i-1)\big)\Bigg).$$
(3.110)

This analysis has been carried out under the assumption that harvesting occurs after mortality and reproduction have taken place (see Figure 2.2). Equation (3.108) can be easily modified if harvesting is assumed to occur before reproduction by changing

$$\prod_{j=1}^{i_0-r+1} [1 - h(t+1-j)]\phi\big(x_0(t+1-i_0)\big)$$

to

$$\prod_{j=1}^{i_0-r} [1 - h(t+1-j)]\phi\big([1 - h(t+1-i_0)]x_0(t+1-i_0)\big),$$

etc. In this case, assuming $i_0 = r$, we obtain Schnute's equations (cf. Schnute, 1985, equations (2.6) and (2.7))

$$J\big(h(t)\big) = h(t)x_0(t)/b,$$

and

$$x_0(t+1) = (1+c)[1-h(t)]sx_0(t)$$
$$- c[1-h(t)][1-h(t-1)]s^2 x_0(t-1)$$
$$+ bw_r \phi\big[\big(1-h(t+1-r)\big)x_0(t+1-r)\big]$$
$$- cbw_r\big(1-h(t)\big)s\phi\big([1-h(t-r)]x_0(t-r)\big).$$

The same approach can be taken in deriving aggregated models for size- rather than age-structured populations. The interested reader is referred to Schnute (1987) for an analysis of aggregated size-structure models in the context of exploiting fish stocks.

3.6.2 Stability

In this section, following Bergh and Getz (in press a), we consider the stability properties of equation (3.106), when $b = 1$ (in which case, as mentioned above, x_0 has the interpretation of a stock-biomass variable) and $w_{r-1} = 0$. Stability results are presented here in terms of the three remaining parameters: $0 < s < 1$, $0 < c < 1$, and $w_r = w > 0$. When $b = 1$ and $w_{r-1} = 0$, equation (3.106) reduces to (dropping the subscript 0 for notational convenience)

$$x(t+1) = (1+c)sx(t) - cs^2 x(t-1) + w\phi\big(x(t+1-r)\big). \quad (3.111)$$

Equation (3.19) and certain special cases thereof are simple enough to obtain analytical results, but complex enough to exhibit cyclic and even "chaotic" behavior (see May and Oster, 1976). The stability properties of the more general equations (3.105) and (3.106) can be investigated using numerical techniques.

If equation (3.111) has an equilibrium solution $\hat{x}$, then $\hat{x}$ clearly satisfies the equation

$$\hat{x} = \frac{w}{1-(1+c)s + cs^2}\phi(\hat{x}).$$

125

As discussed in Section 3.3.2, the specific form of the stock-recruitment function ϕ and specific parameter values will determine the number of equilibrium solutions to equation (3.111). Following the methods outlined in Section 3.1.2, the stability properties of equation (3.111) can be analyzed in a neighborhood of a particular equilibrium solution $\hat{x}$ using linearization techniques. In fact, using $\Delta x(t)$ to denote the perturbation in $x(t)$ from the equilibrium $\hat{x}$ (see equation (3.3)), the linearized version of equation (3.111) is

$$\Delta x(t+1) = (1+c)s\Delta x(t) - cs^2 \Delta x(t-1)$$
$$+ w\hat{d}\Delta x(t+1-r), \qquad (3.112)$$

where

$$\hat{d} = \left.\frac{d\phi}{dx}\right|_{\hat{x}}.$$

A well-known result in systems theory is that the transformation

$$y_1(t) = x(t),$$
$$y_i(t) = y_{i-1}(t-1), \quad i = 2,\ldots,n \qquad (3.113)$$

can be used to transform the scalar n-order difference equation

$$x(t+1) = \sum_{i=1}^{n} a_i x(t+1-i) \qquad (3.114)$$

into the n-dimensional first order difference equation (this is easily shown by direct substitution)

$$\mathbf{y}(t+1) = \begin{pmatrix} a_1 & \cdots & a_{n-1} & a_n \\ 1 & \cdots & 0 & 0 \\ \vdots & \ddots & \vdots & \vdots \\ 0 & \cdots & 1 & 0 \end{pmatrix} \mathbf{y}(t). \qquad (3.115)$$

From Lemma 2.2 it is easily shown that the eigenvalues associated with system (3.115) satisfy the equation (cf. equation (2.35))

$$\sum_{i=1}^{n} \frac{a_i}{\lambda_i} = 1. \qquad (3.116)$$

But since equation (3.114) is equivalent to system (3.115) the eigenvalues satisfying equation (3.116) also determine the stability properties of equation (3.114). Thus the characteristic eigenvalue equation that determines the stability properties of equation (3.111) is simply

$$\lambda^r - (1+c)s\lambda^{r-1} + cs^2\lambda^{r-2} - w\hat{d} = 0. \qquad (3.117)$$

We have already shown in Chapter 2 that the solution to system (3.115) can be expressed in terms of its eigenvalues (cf. equation (2.18)). Thus from the relationship between y and x, as defined by equations (3.113), it follows that the solution to equation (3.111) can also be expressed in terms of these same eigenvalues. Specifically, it can be shown that the solution has the form

$$x(t) = \sum_{i=1}^{r} \kappa_i \lambda_i^t,$$

where κ_i, $i = 1, \ldots, r$, are a set of constants determined from the initial sequence $\{x(1-r), \ldots, x(0)\}$ (since an rth order delay process requires r immediately past consecutive values to calculate the current value).

In the analysis of solutions λ to equation (3.117), it is helpful to consider the graph of the function

$$f(\lambda) = \lambda^r - (1+c)s\lambda^{r-1} + cs^2\lambda^{r-2}$$
$$= \lambda^{r-2}(\lambda - cs)(\lambda - s). \qquad (3.118)$$

Real-valued solutions to equation (3.117) correspond to points where the graph of $f(\lambda)$ is intersected by the line

127

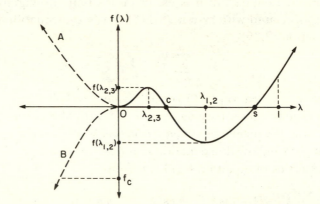

FIGURE 3.3. Graph of the function defined by equation (3.118) when $r > 2$ is even (A) and odd (B).

$f(\lambda) = w\hat{d}$ (Figure 3.3). Some of the following statements relating to expression (3.118) follow directly from Figure 3.3. Others can be derived, as discussed in Bergh and Getz (in press a).

- The equation $f(\lambda) = 0$ has the r solutions $\lambda_1 = s$, $\lambda_2 = cs$, and $\lambda_3 = \ldots = \lambda_r = 0$.
- As $f(\lambda)$ decreases below zero it reaches a value at which the two roots λ_1 and λ_2 coalesce at some point $\lambda_{1,2} \in [cs, s]$.
- Similarly, as $f(\lambda)$ increases above zero it reaches a value at which the two roots λ_2 and λ_3 coalesce at some point $\lambda_{2,3} \in [0, cs]$.
- Note that $\lambda_{1,2}$ and $\lambda_{2,3}$ are solutions to the quadratic term in the equation

$$\frac{df}{d\lambda} = \lambda^{r-3}\left(r\lambda^2 - (1+c)s(r-1)\lambda + cs^2(r-2)\right) = 0,$$

that is,

$$\lambda_{1,2} = \frac{(1+c)s}{2}\left(\frac{r-1}{r}\right)\left(1 + \sqrt{1 - \frac{c}{1+c}\frac{r(r-2)}{(r-1)^2}}\right)$$

128

and

$$\lambda_{2,3} = \frac{(1+c)s}{2} \left(\frac{r-1}{r} \right) \left(1 - \sqrt{1 - \frac{c}{1+c} \frac{r(r-2)}{(r-1)^2}} \right).$$

- For $f(\lambda) > f(\lambda_{1,2})$, the root λ_1 corresponding to the intersection of a horizontal line with the right arm of the graph $f(\lambda)$ is the dominant eigenvalue (analogue of the Perron root for the Leslie matrix equation). Thus solutions to equation (3.112) will either approach or move away from zero (that is, solutions to (3.111) will either approach or move away from $\hat{x}$), respectively, depending on whether $\lambda_{1,2} < \lambda_1 < 1$ or $\lambda_1 > 1$. The approach to or from $\hat{x}$ will become monotonic as $t \to \infty$, although, at least for $r \geq 4$ some oscillations will be evident for small t from the subdominate negative root (r even) and/or the complex conjugate roots that bifurcate[3] from $\lambda = 0$ as $f(\lambda)$ increases from zero.
- When $\lambda_1 = 1$, it follows from equation (3.118) that

$$f(1) = (1 - cs)(1 - s),$$

which may or may not exceed the value of $\lambda_{2,3}$ given above, depending on the particular values of c, s, and r.
- There exists a value $f_c < f(\lambda_{1,3})$ at which the complex conjugate pair of eigenvalues λ_1 and λ_2, bifurcating at the value $\lambda_{1,2}$ lie on the unit circle (have modulus 1). For $f_c < f(\lambda) < f(\lambda_{1,2})$, the complex conjugate pair of eigenvalues λ_1 and λ_2 that are solutions to (3.117) are the dominant eigenvalues with modulus less than 1 (specifically, their modulus lies on the interval $(\lambda_{1,2}, 1)$). Thus solutions to equation (3.112) will oscillate around, but steadily approach the equilibrium $\hat{x}$, for $f(\lambda) \in [f_c, f(\lambda_{1,2})]$.

[3] Opposite of coalesce.

- For $f(\lambda) < f_c$, the complex conjugate pair of eigenvalues λ_1 and λ_2 lie outside the unit circle and the equilibrium is unstable with solutions to the linearized equation (3.112) oscillating about zero or, equivalently, solutions to equation (3.111) oscillating around $\hat{x}$.

For the linearized system (3.112), the period of oscillation corresponding to a negative root is 2. The period of oscillation corresponding to a pair of complex conjugate roots depends on the location of these roots in the complex plain (for details, see Levin and May, 1976; Bergh and Getz, in press a). Even for very simple models involving one time delay ($c = 0$ and $r = 1$ in equation (3.111)) it is known that the period of oscillation can increase to ∞, at which point the behavior of the solution is termed chaotic because the solution oscillates around the equilibrium $\hat{x}$ in an aperiodic and, hence, seemingly random fashion (also see May and Oster, 1976).

3.7 REVIEW

At the beginning of this chapter we took a first step towards extending the linear matrix age- and stage-structured models of the previous chapter to nonlinear models by modifying the linear first-year survival relationship

$$x_1(t + 1) = s_0 x_0(t)$$

to

$$x_1(t + 1) = s_0 x_0(t)\psi\big(x_0(t)\big).$$

The "modifier" function $\psi(x_0)$ typically decreases from unity as the density of newborns x_0 increases. Since each breeding individual has the same average natality, the variable x_0 (after multiplying by a constant or, equivalently, changing units) can also be interpreted as a breeding-stock variable. With this interpretation the first-year survival relationship becomes a breeding-stock/new-recruits relationship or, more simply, a stock-recruitment relationship. In general, one can relate the stock r-years-ago to the number of r-year-olds now and obtain the relationship

$$x_r(t + r) = \phi\big(x_0(t)\big),$$

thereby lumping all the effects of mortality in the first r years of an individual's life into a single function ϕ. This only has value in a management setting if harvesting or other management actions do not impact individuals less than r years old.

The association of density-dependent survival with one or more of the population stage classes fundamentally alters the properties of the modified linear matrix model: an equilibrium population level may now exist as a consequence of density dependence acting to decrease per capita population growth below replacement levels (mortality exceeds birth) at high population densities. Density dependence will not always guarantee the existence of an equilibrium; but, since no biological population can grow without bound, all biological populations that increase at low densities will have one or more equilibrium densities. A population will grow at low densities only if its stock value $R_0 = \sum_{i=1}^{n} b_i l_i$ is greater than 1. Populations for which $R_0 < 1$ are unable to fully replace their numbers each generation, even at low densities; such populations eventually die out.

It is clear, from the expressions derived for the population equilibrium densities associated with the different stock-recruitment functions that these densities are determined by the stock value R_0 and the form of the density-dependent survival functions. The latter, however, is more critical than the stock value when it comes to determining the stability properties of these equilibrium densities. For example, if an equilibrium exists in the case

131

of the Beverton and Holt stock-recruitment relationship, then it is always stable; that is, the stock value has no influence on the stability properties of the equilibrium other than determining whether an equilibrium exists.

In our nonlinear treatment of harvesting in this chapter, the same kind of problems are formulated as in our linear treatment in the previous chapter. Now, however, if a population has an equilibrium density when unharvested, then it will also have an equilibrium density when harvested, providing not too many individuals are harvested from any one stage class. The value of this equilibrium will depend on the harvest policy, and at least one of these policies will produce a maximum yield which we refer to as the maximum sustainable yield (MSY). The MSY policy corresponds to specifying the number of individuals that must be removed from each age or stage class (which applies in each time period) rather than just an optimal proportion, as obtained by solving the linear problem (without population size constraints).

In the first part of the chapter, the density-dependent behavior in the population is assumed to be a function of a variable that is aggregated across stage classes, namely, breeding stock. The idea that density dependence can be more succinctly expressed in terms of aggregated variables than in terms of the individual state variables themselves is pursued in the second part of this chapter. This approach allows us to preserve much of the linear structure of the problem and use this to advantage in solving for equilibrium population density levels associated with a specific model. In particular, if m aggregated variables are introduced, then finding the equilibrium involves solving m simultaneous nonlinear equations in the m aggregated variables. In contrast, finding the equilibria corresponding to a general nonlinear model involves solving as many simultaneous equations as there are age or stage classes in the model.

The aggregated modeling approach also facilitates the analysis of the stability properties of the equilibria, because linearization of the model around these equilibria involves fewer derivatives if m is small. Finally, the aggregated modeling approach leads to clearer understanding of how nonlinearities affect the form of the optimal harvesting policy. Specifically, each additional aggregated variable that enters the model nonlinearly implies that at most one additional age or stage class in the population must be harvested to obtain the MSY solution. In the absence of migration, the existence of one aggregated variable implies that individuals from at most two stage classes must be harvested to obtain the MSY; or two aggregated variables implies harvesting three stage classes, etc.

MSY solutions are useful in providing a reference point for solutions to the general problem of managing populations not in equilibrium. Nonequilibrium problems arise because a population is either not initially at the density corresponding to an MSY solution, or the population is influenced by stochastic phenomena and is constantly deviating from MSY levels. In the next chapter, we show how the MSY conditions play a central role in solutions to dynamic harvesting problems.

One of the tools that can be used to find solutions to dynamic harvesting problems is Pontryagin's Maximum Principle. This principle is most often stated for problems associated with continuous time models (that is, differential equation models). A discrete version of the principle is required for the discrete-time models presented here. The principle is used to discuss the problem of determining endpoint boundary conditions for problems associated with finite planning horizons. Some of the numerical procedures used to solve harvesting problems presented in the next two chapters also make use of Pontryagin's Maximum Principle, but these procedures are not discussed in detail. A primary purpose for presenting the principle, as we have here, in a from that is applicable to resource management problems is that such presentations are not readily available in the resource or population management literature.

There is a fundamental difficulty, in developing stochastic models for nonlinear systems, that does not arise with linear systems: there is no certainty-equivalence principle that allows us to use the deterministic model as a description for the dynamics of stage-class means; and, it is not possible to directly generate equations for the variance/covariance terms using Kronecker products. In fact, the equations for the means involves variance/covariance terms, while equations for the variance/covariance terms involve third-order moments. In general, there is no way of getting an exact description of how the first- and second-order moments of the stage-class distributions evolve without making some type of approximation. We discussed two types of approximations. In the first we assumed that the stochastic effects are relatively small so that third- and higher-order moments around the mean can be neglected. In the second we assumed that the stage-class distributions have some predetermined two-parameter form (e.g., lognormal, gamma), and then found equations to solve for these distributions.

The equations obtained using both these approaches allow us to predict directly how the first two moments of the stage-class density distributions evolve with time. It is not always apparent how well these approximate equations for the means and variance/covariance terms predict work. This is especially true if the nonlinearities in the model are quite complex. In this case, Monte Carlo simulation techniques can be used to directly simulate the behavior of stochastic resource systems. In this context, complex nonlinearities do not present a problem since all the work is numerical (e.g., no derivatives need to be evaluated). The drawback of the Monte Carlo method is the number

of times a particular simulation must be repeated using a
random number generator, so that sufficient possible solutions are generated to obtain statistics relating to the
most probable solutions (of course we are assuming that
the random number generator itself is generating numbers that appear to be random). Once a sufficient number
of solutions has been generated, the Monte Carlo method
is powerful because such questions as "what is the probability that the stock biomass will fall below a given level?"
can be answered directly from the simulation data. Examples of this type of analysis are discussed in the next
chapter.

In the final section of this chapter we demonstrated that
lumped-variable and stage-structured models are equivalent if we are prepared to make a number of simplifying assumptions, including that survival rates are constant
with age, and that, after maturity, fecundity is constant or
directly related to size. These assumptions dramatically
reduce the number of parameters in the model. This is a
distinct advantage if, as happens in fisheries, we are trying to estimate the value of these parameters from observations or inferences on the density of the population at
different points in time.

CHAPTER FOUR

Fisheries Management

4.1 BACKGROUND

The exploitation of fisheries is a multibillion dollar industry in countries like the United States and Japan. Although there is still great potential to exploit the productivity of the sea by harvesting species like Bering sea pollock (*Theragra chalcogramma*) in the northern hemisphere and antarctic krill (*Euphausia superba*) in the southern hemisphere, a number of highly productive species, notably Peruvian anchoveta (*Engraulis ringens*), have been fished to exhaustion (at least for the foreseeable future). In an effort to improve the efficiency of exploiting stocks and prevent the collapse of key fisheries, a concerted effort was made in the 1950s to develop a quantitative theory of fisheries management. Notable is the work of Ricker (1954) that has led to the development of so-called surplus yield models and the work of Beverton and Holt (1957) who laid the foundation for "cohort analysis."

The surplus yield models view fisheries as lumped resources, which are typically represented by a scalar biomass variable whose dynamics are modeled by a single nonlinear differential or difference equation. The scalar characterization of the problem makes it considerably more amenable to mathematical analysis than some vector characterizations like cohort analysis. A comprehensive economic theory of resource exploitation has developed around the surplus yield model, primarily by Clark and his collaborators (see Clark, 1976, 1985).

Here we will focus on Beverton and Holt's cohort approach and the age-structured models that have developed from it. In the Beverton and Holt theory it is assumed

136

that the number of new individuals recruited to the fishery each year is constant. Because of the multidimensional nature of the cohort formulation, a general theory of exploitation is much more difficult to develop if a nonlinear stock-recruitment relationship is assumed. A linear stock-recruitment relationship leads back to Leslie matrix theory (see Chapter 2). Here we will use the methods presented in Chapter 3 to develop a theory for exploiting age-structured marine fisheries exhibiting nonlinear recruitment.

In the next section we review Beverton and Holt's cohort analysis and then extend the theory to incorporate nonlinear recruitment. In three subsequent sections we go on to discuss exploitation under equilibrium, nonequilibrium, and stochastic conditions. Finally, in the last two sections we take a look at fisheries which respectively involve several species and several participants (exploiters).

4.2 DETERMINISTIC MODELS

4.2.1 Cohort Model

In this chapter we treat t as a continuous variable of time. Thus, only in this chapter, will we use k to denote discrete integer values of time. The cohort model, also commonly know as the "dynamic pool" model, considers the dynamics of a cohort subjected to a constant natural mortality rate α (M in the fisheries literature) from the time of its recruitment to the fishery at age r or, more generally, at some age $t_v \geq$, and a constant fishing mortality rate v (F in the fisheries literature). Suppose that x_0 is the number of individuals born at time $t = 0$, and that $x(t)$ is the number of individuals still in this cohort at time t. Then $x(t)$ satisfies the differential equation

$$\frac{dx}{dt} = -\alpha x, \qquad 0 \leq t \leq t_c$$

$$= -(\alpha + v)x, \qquad t \geq t_c \qquad (4.1)$$

and

$$x(0) = x_0.$$

The solution to this equation is simply

$$x(t) = \begin{cases} x_0 e^{-\alpha t}, & 0 \leq t < t_c; \\ x_0 e^{-\alpha t_c - v(t - t_c)}, & t \geq t_c. \end{cases} \qquad (4.2)$$

Let $w(t)$ be the weight of an individual fish at age t. The weight function $w(t)$ is often assumed to satisfy the following assumptions:

1. $\frac{dw(t)}{dt} > 0$ for all $t \geq 0$;
2. $w_0 < w(t) < w_\infty$ for all $t > 0$;
3. $\frac{d}{dt}[\frac{1}{w(t)} \frac{dw(t)}{dt}] < 0$ for all $t \geq 0$.

Assumptions 1 and 2 imply that an individual's age is always increasing from an initial level w_0, but is bounded above and therefore asymptotically approaches some upper value w_∞. Assumption 3 implies that the proportional rate of increase decreases with age. These three assumptions appear to be satisfied by most species of fish. In fact, weight is often fitted to a function that satisfies these assumptions. This function is referred to as the von Bertalanffy growth curve and has the three-parameter form

$$w(t) = \bar{w}(1 - c_1 e^{-c_2 t})^{c_3}, \qquad (4.3)$$

where the constants c_i are positive, and c_3 typically lies between 2.5 and 3. A value of c_3 close to 3 reflects the fact that an individual's growth is approximately a saturating exponential function $(1 - c_1 e^{-c_2 t})$ in each of its three linear dimensions (i.e., width, length, height).

If all fish are assumed to be exactly alike, then the total "biomass" $B_{r,v}(t)$ of a cohort born at time $t = 0$ and subject to a fishing mortality rate v after age r is

$$B_{r,v}(t) = w(t)x(t), \qquad t \geq 0. \qquad (4.4)$$

Note from equations (4.2) and (4.4) that, in the absence of fishing ($v = 0$),

$$B_{r,0}(t) = w(t)x_0 e^{-\alpha t}, \qquad t \geq 0,$$

which has an extremum (which we show is a maximum) at time $\hat{t}$ when the derivative of $B_{r,0}(t)$ is zero. Differentiating $B_{r,0}(t)$ we obtain for $t \geq 0$

$$\frac{dB_{r,0}}{dt} = w(t)x_0 e^{-\alpha t} \left(\frac{1}{w(t)} \frac{dw(t)}{dt} - \alpha \right).$$

Thus by Assumptions 1 to 3, it follows that if an extremum occurs at $\hat{t}$, it is a maximum and the following equation is satisfied:

$$\frac{1}{w(\hat{t})} \frac{dw(\hat{t})}{dt} = \alpha. \tag{4.5}$$

If we were not concerned with allowing individuals to live beyond age $\hat{t}$ for the purposes of reproduction, and we had no upper limit on the fishing mortality rate v, then it is clear that we obtain maximized yield from a cohort by harvesting every individual the moment it reaches age $\hat{t}$. This follows since after time $\hat{t}$ more biomass is lost from the cohort due to natural mortality than is gained due to the growth of individuals (see equation (4.5)).

There are three major factors, however, that make this an unworkable solution. First, many stocks consist of a mixture of several cohorts, each born at a different time k, and there is no fishing gear that can precisely select (for example, through choice of appropriate mesh size) fish of a certain age and above. Second, if individuals reach sexual maturity at an age greater than $\hat{t}$, then a proportion of individuals must be left to contribute to reproduction. Thus even if we could harvest by age, there is a trade-off between maximizing the expected biomass of each individual and allowing adequate time for individuals to reproduce. This in essence explains why bimodal harvesting

139

policies are required to maximize yield in age-structured systems that have a relationship between stock and recruitment (see Chapter 3). In the cohort model, recruitment is independent of stock and the maximum yield policy is unimodal (harvest all individuals at $\hat{t}$). Third, there will always be a limit to the level at which fishing effort can be applied.

These limitations will be dealt with below, but in terms of equation (4.1) it is clear that the biomass yield obtained from a cohort born at time $t = k$, fished at a constant mortality rate v over the total time it spends in the fishery is

$$Y(v, t_c) = \int_{t_c}^{\infty} vw(t)x(t)dt. \tag{4.6}$$

Thus it follows from (4.2) that

$$Y(v, t_c) = vx_0 e^{\alpha t_c} \int_{t_c}^{\infty} w(t)e^{-(\alpha+v)t}dt. \tag{4.7}$$

Beverton and Holt pointed out that the quantity $Y(v, t_c)$ expressed in (4.7) has another interpretation. Suppose that a new cohort is spawned every year at the same level x_0, and that each of these cohorts is harvested from age t_c onwards using a constant fishing mortality level v. Then $Y(v, t_c)$ is also the total biomass yield obtained from a population that consists of all cohorts that are simultaneously harvested over a single time period; that is, $Y(v, t_c)$ is the annual equilibrium biomass, provided x_0 is the equilibrium recruitment level that corresponds to the harvesting parameters v and t_c. This equivalence of harvesting a single cohort over an infinite time horizon and harvesting an equilibrium assemblage of all cohorts over a single period of time follows because

$$vx_0 e^{vt_c} \int_i^{i+1} w(t)e^{-(\alpha+v)t}dt$$

is the harvest obtained from the ith age class in the assemblage, but it also is the current (yearly) harvest of individuals from the cohort born i years prior to the year in which the harvest was taken.

Under a particular harvesting policy (v, t_c), the quantity $Y(v, t_c)/x_0$ is referred to as the *yield per recruit*. Knowing this quantity for all policies (v, t_c) allows one to answer the question (see Gulland, 1983), "What pattern of fishing will give the greatest yield from the year-class of fish that has just been recruited?" If one assumes that recruitment is a constant x_0, independent of stock size, then the *yield per recruit isopleths* or level contours $Y(v, t_c)/x_0 = c$, where c is an arbitrary constant, are obtained directly from expression (4.7) (Figure 4.1). Note that from our definition of $\hat{t}$ as the age at which the biomass of each cohort is maximized, it follows that the maximum value of $Y(\infty, t_c)/x_0$ is achieved at $t_c = \hat{t}$; that is, when the fishing effort level is infinitely high, the total cohort is harvested at a single point in time, whence the maximum biomass is obtained at $t = \hat{t}$.

4.2.2 Seasonal Harvesting with Nonlinear Recruitment

Beverton and Holt's cohort analysis is unable to deal with fisheries in which recruitment is dependent upon the level of the stock. In most fisheries there is a direct relationship between the number of eggs spawned and the number of spawning females; thus there must exist an underlying stock-recruitment relationship. This relationship is often masked by stochastic environmental influences and may be apparent only at very low stock densities when, for example, poor recruitment is evident even under favorable environmental conditions. Thus the yield per recruit concept has value when recruitment is essentially constant on average over the range of stock sizes that support an economically viable fishery. It is also useful when recruitment of large cohorts is detected using,

141

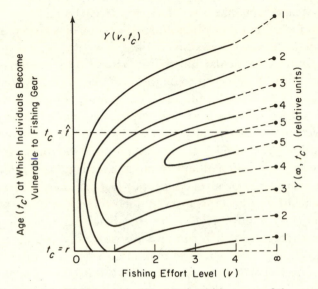

FIGURE 4.1. Yield isopleths $Y(v, t_c)$ plotted in terms of the age-selectivity parameter $t_c \geq r$ (r is age at recruitment) and the fishing effort level parameter v. (Adapted from Beverton and Holt, 1957.)

say, larval survey techniques, and an assessment is needed on how best to exploit this cohort as it passes through the fishery. Of course, the value of this approach is diminished in most fisheries by the inability of fisherman to target their fishing effort on a particular cohort.

The form of the underlying stock-recruitment relationship obviously will influence the mean dynamics of a fish stock over the long term, as short-term fluctuations are averaged out. Thus an estimation of the form of the stock-recruitment relation is essential in assessing the long-term productivity of a fishery. This is much easier said than done since estimating the stock and recruitment levels in any particular year, using survey data or subsequent *catch-at-age per unit effort* data, poses enormous problems for fisheries scientists (see Gulland, 1983; Pope and Shepherd, 1982, 1985; Shepherd and Nicholson, 1986; Walters, 1986). The question of estimation is an extremely important one, but it is not the subject of this book. Here we always assume that the parameters used in our harvesting models have already been estimated from raw data. During the operation of a fishery, new data should always be incorporated to improve the estimates of the population parameters (see Walters, 1986). Questions relating to the stochastic nature of the stock-recruitment process will be dealt with in a subsequent section. In this section, we analyze the exploitation of stocks that are subject to a known deterministic stock-recruitment process.

The cohort analysis of the previous section assumes that the gear used in the fishery exhibits *knife-edge* selectivity (Figure 4.2). More generally, the vulnerability of individuals varies with age. For example, the smaller the fish, the more easily it will escape from a fishing net of a given mesh size. If v is a measure of the *intensity* of fishing (e.g., number of boats, nets, or hooks), then the number of fish, u_i, harvested from age class i over the period $[k, k+1)$ is determined by $u_i = q_i v(k) \int_k^{k+1} x_i(t)dt$, where q_i is termed the *catchability coefficient*. By appropriately scaling the units of v, the catchability coefficients can always be normalized (in which case they are often referred to as *selectivity coefficients*) so that the maximum catchability coefficient has the value 1. The elements q_i determine a gear *selectivity ogive* (Figure 4.2).

A number of different assumptions can be made with respect to the time that spawning actually occurs and when the fishing season opens and closes. We assume, for our treatment here, that spawning occurs at the beginning of each time interval $[k, k+1]$, at which time harvesting begins at intensity $v(k)$ and remains at this intensity for the length of the harvesting period $(k, k+\bar{t})$ (open season). A

143

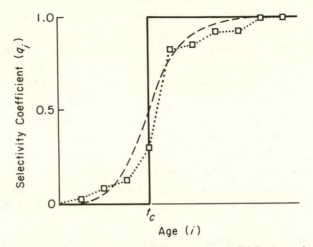

FIGURE 4.2. Knife-edge selectivity at age t_c (solid line), a selectivity ogive characterized by its slope and the age t_c at which the catchability coefficient is 0.5 (broken line), and an empirical selectivity ogive (dotted line).

similar analysis can be carried out if spawning occurs at some other time on the interval $[k, k + 1]$.

For generality, we assume that the natural mortality rate is age dependent; that is, the ith age class experiences a natural mortality rate α_i. Thus if $x_i(t)$ is the number of individuals in the ith age class at time t, then it follows, as in deriving equation (4.1), for $i = 1, 2, \ldots$, that

$$\frac{dx_i}{dt} = -\left(\alpha_i + q_i v(k) x_i\right), \qquad t \in [k, k + \bar{t}]$$
$$= -\alpha_i x_i, \qquad t \in (k + \bar{t}, k + 1). \tag{4.8}$$

Integrating equations (4.8) over $[k, k + 1)$ and accounting for the transition to an older age class, we obtain the transition equations

$$x_{i+1}(k + 1) = e^{-\left(\alpha_i + q_i v(k) \bar{t}\right)} x_i(k), \qquad i = 1, 2, \ldots. \tag{4.9}$$

144

To complete the description of the transition process, we add a general stock recruitment relationship, as discussed in Chapter 3; that is (cf. equation (3.8)),

$$x_r(k+r) = l_r x_0(k) \psi\big(x_0(k)\big), \qquad (4.10)$$

where x_0 is interpreted as an aggregated egg variable or as the number of newborns (as in Beverton and Holt's cohort model), or as the spawning stock biomass variable (when spawning stock biomass is correlated with the number of eggs laid). Thus for a given set of fecundity-related coefficients b_i, x_0 is expressed as

$$x_0(k) = \sum_{i=1}^{n} b_i(k) x_i(k). \qquad (4.11)$$

Note that equation (4.9) should be truncated at age n, if no or very few individuals live beyond age n. Otherwise, a more appropriate approximation is

$$x_n(k+1) = e^{-\left(\alpha_{n-1}+q_{n-1}v(k)\hat{t}\right)} x_{n-1}(k)$$
$$+ e^{-\left(\alpha_n + q_n v(k)\hat{t}\right)} x_n(k). \qquad (4.12)$$

Consider the system of equations (4.9) for $i = r, \ldots, n-1$, (4.10), (4.11), and (4.12). If recruitment is taken as a constant (that is, independent of x_0), then these equations represent the extension of a cohort model to seasonal harvesting with a general rather than knife-edge selectivity ogive. If recruitment is linear, then these equations are equivalent to a Leslie matrix model (cf. Chapter 2) with survival rates

$$s_i = e^{-\alpha_i} \qquad (4.13)$$

and proportional harvesting rates

$$h_i(k) = 1 - e^{-q_i v(k)\hat{t}}. \qquad (4.14)$$

In contrast to the Leslie matrix harvesting problem, however, the proportions h_i cannot be chosen independently since they all depend on the choice of the scalar harvesting effort variable $v(k)$. This linear model is invariably inappropriate for fisheries analysis (the constant recruitment cohort model being closer to the truth except at relatively low stock levels) and one of the stock-recruitment relationships (3.9) to (3.14) should be selected to represent the dynamics of the stock.

It follows from equation (4.8) that the yield obtained during the $(k + 1)$th time period is (cf. expression (4.6))

$$Y_k(v(k), \mathbf{q}) = \int_k^{k+1} v(k) \sum_{i=1}^{n} q_i w_i(t) x_i(t) dt, \qquad (4.15)$$

where $\mathbf{q}$ is the vector of catchability coefficients and $w_i(t)$ (the biomass of an individual aged i at time $t \in [k, k + 1]$) can be determined from the general growth function $w(t)$ (e.g., function (4.3)) using the relationship

$$w_i(t) = w(t - k + i). \qquad (4.16)$$

4.3   EQUILIBRIUM YIELD ANALYSES

4.3.1   Yield Effort Curves

For the cohort analysis we discussed the concept of yield per recruit as a function of the harvesting policy variables (v, t_c). In the general stock-recruitment case, total yield is more informative than yield per recruit since the equilibrium recruitment value changes as a function of the harvesting effort variable v.

The equilibrium recruitment level, $\hat{x}_1$, and stock index, $\hat{x}_0$, corresponding to an effort level $\hat{v}$ are easily calculated using the theory developed in Chapter 3. Specifically, the fisheries model developed in the previous section is a proportionally harvested age-structured population for which the parameters l_i and ν_i in the equilibrium age-class relationship (cf. equation (3.40)) $x_i = l_i\nu_i x_0 \psi(x_0)$ is given by (see equations (2.2), (3.38), (4.13), and (4.14))

$$l_i = e^{-\sum_{j=0}^{i} \alpha_j},$$

$$\nu_i(\hat{v}) = e^{-\hat{v}t \sum_{j=1}^{i} q_i}.$$

Note that $s_n = e^{-\alpha_n} > 0$ so the l_{n-1} has the modified form

$$l_{n-1} = e^{-\sum_{j=0}^{n-1} \alpha_j} \bigg/ \left(1 - e^{-(\alpha_n + q_n \hat{v}t)}\right),$$

which now is dependent on the value of $\hat{v}$. The equilibrium recruitment level $\hat{x}_1$ can now be calculated using equations (3.41) and (3.42). Further, if such an equilibrium exists (i.e., the solution to these equations is nonnegative) then the corresponding stock level, x_0, is given by (cf. equation (3.21) and recall that $R_0 = \sum_{i=1}^{n} b_i l_i$)

$$\hat{x}_0 = \left(\sum_{i=1}^{n} b_i l_i \nu_i\right) \frac{\hat{x}_1}{s_0}. \tag{4.17}$$

Furthermore, the stability properties of the equilibrium corresponding to $\hat{v}$ can be evaluated following the methods outlined in Chapter 3.

Once x_0 is known, then all the values x_i can be generated using equation (3.40), and the equilibrium yield $Y(\hat{v}, \mathbf{q})$ can be obtained from expression (4.15). Since the equilibrium solution is independent of the value of k and $\hat{x}_i(t) = \hat{x}_i e^{-(\alpha_i + q_i \hat{v})t}$ for $t \in [0, \bar{t}]$, it follows that expression (4.15) becomes

$$Y(\hat{v}, \mathbf{q}) = \hat{v} \sum_{i=1}^{n} q_i \hat{x}_i \int_0^{\bar{t}} w_i(t) e^{-(\alpha_i + q_i \hat{v})t} dt. \tag{4.18}$$

This expression can be integrated if the form of $w_i(t)$ is known.

In the simplest case where $w_i(t)$ is a set of empirically determined constants w_i, expression (4.18) reduces to

$$Y(\hat{v}, \mathbf{q}) = \hat{v} \sum_{i=1}^{n} w_i q_i \hat{x}_i (1 - e^{-(\alpha_i + q_i \hat{v})\hat{t}})/(\alpha_i + q_i \hat{v}). \quad (4.19)$$

If $w_i(t)$ is derived, using identity (4.16), from a von Bertalanffy growth function (4.3) for which the constant $c_3 = 3$, then (4.18) is easily integrated to obtain

$$Y(\hat{v}, \mathbf{q}) = \hat{v}\bar{w} \sum_{i=1}^{n} q_i \hat{x}_i$$

$$\times \sum_{j=1}^{3} \binom{3}{j} \frac{(-1)^j c_1^j e^{-jc_2 i}(1 - e^{-(jc_2 + \alpha_i + q_i \hat{v})\hat{t}})}{jc_2 + \alpha_i + q_i \hat{v}},$$

where $\binom{3}{j} = \frac{3!}{j!(3-j)!}$ are the binomial coefficients in the expansion of a cubic.

4.3.2   Optimal Yields

In a trawl fishery where the nets are characterized by mesh diameter or a line fishery where the hooks are also characterized by a scalar size measurement, we can think of the gear as being parameterized by t_c (cf. Figure 4.2) even though selection is not knife edge. Thus the selectivity coefficients can be represented by $\mathbf{q} = \mathbf{q}(t_c)$, and yield, as in expression (4.7), is a function of v and t_c. If we now examine yield isopleths in (v, t_c) space, an important change occurs compared with the plots obtained for the cohort model illustrated in Figure 4.1. We no longer have the property that $Y(v, \mathbf{q}(t_c))$ is an increasing function of v for given t_c. The reason for this is that, depending on the particular stock-recruitment relationship, recruitment may well decrease when the stock is severely impacted to the point where a positive yield level can no longer be sustained (an equilibrium does not exist). This situation is illustrated in Figure 4.3 below with respect to the South African anchovy (*Engraulis capensis*) fishery.

In terms of cohort analysis Beverton and Holt (1957) introduced the concept of the *eumetric* yield curve and, in the same context, Clark (1985) discusses the dual concept of the *cacometric* yield curves. These concepts apply equally well to the more general nonlinear recruitment, one parameter selectivity ogive discussed here. Specifically, for any fixed level of fishing effort $\hat{v}$, there is an optimal gear type characterized by $\mathbf{q}(t_c)$, such that $Y(\hat{v}, \mathbf{q}(t_c))$ is maximized with respect to t_c. The resulting yield/effort curve is called the eumetric yield and is formally defined by

$$Y_{\text{eumetric}}(v) = \max_{t_c} Y(v, \mathbf{q}(t_c)).$$

The cacometric yield curve is the maximum over effort for given t_c and is formally defined by

$$Y_{\text{cacometric}}(t_c) = \max_{v} Y(v, \mathbf{q}(t_c)).$$

The above two yield curves are just the projection of the envelope curves of $Y(v, \mathbf{q}(t_c))$ onto the v and t_c axes, respectively (e.g., see Figure 4.4 below).

Once either of these two yield curves has been constructed, the point (v^*, t_c^*) that maximizes $Y(v, \mathbf{q}(t_c))$ can be determined. The value of these parameters and the corresponding value Y^* are not the only quantities of interest. Since it is not always possible to precisely regulate the value v in most fisheries, or accurately characterize $\mathbf{q}$ in terms of t_c, it is also important to evaluate how flat the surface $Y(v, \mathbf{q}(t_c))$ is in the region of Y^*, that is, how sensitive the maximum yield is to changes in v and t_c. Furthermore, evaluation of $Y(v, \mathbf{q}(t_c))$ ignores the cost associated with the implementation of a particular effort level v and all other economic considerations. Fisheries managers often look at *yield per unit effort* as a means of incorporating the cost of fishing into their management analyses. Also the concept of $v^{0.1}(t_c)$ has been introduced for those fisheries where the maximum yield is obtained at $v = \infty$. The quantity $v^{0.1}(t_c)$ is defined for each t_c to be the value of v that corresponds to the effort level where the slope of the yield curve is 10% of its value at the origin. This concept, although used frequently by fisheries managers, is entirely *ad hoc*. A fuller discussion on the economics of exploiting yield can be found in Clark (1976, 1985).

4.3.3 Ultimate Sustainable Yield

In Chapter 3 we posed a number of formal optimization problems and discussed various techniques for finding their solution. Invariably, mathematical formulations are highly simplified abstractions of real problems so that, for one reason or another, the optimal solution to the formally stated problem may be undesirable in practice. (For example, in a dynamic problem the optimal level of effort may change drastically from one season to the next.) A suboptimal solution that circumvents these features may be more desirable if its corresponding yield value is sufficiently close to the formal optimal solution. If the optimal solution is known, then the "loss" involved when implementing a suboptimal but more pragmatic solution can be evaluated and compared with "hidden" costs that become apparent when implementing the optimal solution in the real system.

As already mentioned, the essential difference between the system modeled by equations (4.9) to (4.12) and systems modeled by (3.48) is that the latter implies we have the facility to harvest each age class independently. System (3.48) represents the *ultimate* level of control because each age class can be independently harvested. It is informative to evaluate differences between yields obtained with this ultimate level of control and yields associated with a more restrictive level of control (in systems (4.9) to (4.12)

150

we are able to select only mesh size (t_c) and fishing effort (v)). Specifically, if Y° denotes the *ultimate sustainable yield*, then we are interested in what percentage Y^* (maximum for restricted system) or even $Y(v, t_c)$ for particular (v, t_c) is of Y°.

The ultimate sustainable yield can be defined as the solution to the problem of maximizing the yield subject to the intraseasonal dynamic constraint equation (cf. (4.8))

$$\frac{dx_i}{dt} = -\big(\alpha_i + v_i(t)\big)x_i, \qquad t \in [k, k+1) \qquad (4.20)$$

and the transition period equilibrium constraint $\mathbf{x}(k+1) = \mathbf{x}(k)$. The optimal effort solution $\mathbf{v}^\circ(t) = (v_1^\circ(t), \ldots, v_n^\circ(t))'$ is easily found since it is known to satisfy the following properties (Getz, 1979, 1980a). At most two $v_i^\circ(t)$ are nonzero for $i = 1 \ldots, n$ (bimodality property) and they take the form of impulse functions (infinite value applied at a single point in time) applied at time t_i°, where (cf. (4.5)) t_i° satisfies either the equation

$$\frac{1}{w_i(t_i^\circ)} \frac{dw_i(t_i^\circ)}{dt} = \alpha_i$$

or

$$t_i^\circ = \begin{cases} 0, & \text{if } \frac{1}{w_i(0)} \frac{dw_i(0)}{dt} < \alpha_i; \\ 1^-, & \text{if } \frac{1}{w_i(1)} \frac{dw_i(1)}{dt} > \alpha_i. \end{cases}$$

Note that 1^- is used to denote the fact that the impulse v_i is applied at the end of the time period just prior to transition to the next age class. Applying an infinite level of v_i at time t_i° is equivalent to removing a given number of individuals u_i (not necessarily all the available i-year-olds) at time t_i°; that is, equation (4.20) can be integrated to obtain

$$x_{i+1}(k+1) = e^{-\alpha_i}x_i(k) - e^{-\alpha_i(1-t_i^\circ)}u_i, \quad i = 1, \ldots, n-2,$$
$$x_n(k+1) = e^{-\alpha_n}x_n(k) + e^{-\alpha_{n-1}}x_{n-1}(k)$$
$$- e^{-\alpha_n(1-t_n^\circ)}u_n - e^{-\alpha_{n-1}(1-t_{n-1}^\circ)}u_{n-1}. \quad (4.21)$$

151

As described in Chapter 3, the maximum sustainable yield problem associated with system (4.21) reduces to a linear programming problem in which at most two of the elements of the harvest vector **u** are nonzero (application of Theorem 3.2).

4.3.4 Anchovy and Cod Fisheries

For the purposes of illustration, we review results obtained from a sustainable yield analysis of the South African anchovy fishery (Getz, 1980a,b) and the Arcto-Norwegian cod (*Gadus morhua* Linnaeus) fishery (Reed, 1980).

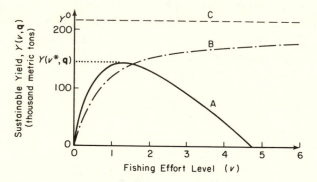

FIGURE 4.3. Sustainable yield levels $Y(v, \mathbf{q})$ for the South African anchovy fishery: A, using the parameters listed in Table 4.1; B, harvesting only 3-years and older individuals; C, the ultimate sustainable yield Y°.

The parameters used to construct yield-effort curves for the anchovy fishery are given in Table 4.1. More recent estimates of these parameters can be found in Bergh and Butterworth (1987). The sustainable yield effort curves obtained for this fishery are illustrated in Figure 4.3. The ultimate sustainable yield is around 210,000 metric tons. It is obtained by harvesting 32% of the individuals in age

TABLE 4.1. Parameter values for the South African anchovy (*Engraulis capensis*) fishery.

Number of age-classes	$n = 5$
Length of harvesting season	$\bar{t} = 2/3$ (years)
Natural mortality parameters	$\alpha_i = 0.8, i = 1, \ldots, 5$
Growth function	$w(t) = 35(1 - 0.73e^{-0.43t})^3$ (g)
Relative fecundity values[a]	$b_1 = 0.5, b_i = 1, i = 2, \ldots, 5$
Stock-recruitment function	$x_1 = \frac{122 \times 10^9 x_0}{11 \times 10^5 + x_0}$ (individuals)
Relative catchability coefficients[b]	$q_1 = 0.24, q_2 = 0.36, q_3 = 0.42, q_4 = q_5 = 1$

[a] b_i in equation (4.17).

[b] These correspond to a purse-seine 13mm net that is currently used in the fishery. For comparative purposes, the yield-effort curves were also obtained when setting $q_1 = 0$, and $q_1 = q_2 = 0$ (see Figure 4.5).

class 3 and 100% of the individuals in age class 5 at the beginning of the harvesting interval. The practice of using 13 mm purse-seine nets, which corresponds to the selectivity ogive listed in Table 4.1, allows an MSY of about 140,000 metric tons (67% of the USY) at $v = 1.2$ (that is, $q_i v$ is 1.5 times the natural mortality rate for the four- and five-year-olds). It is clear from the other yield-effort curves that a larger net, if it essentially does not impact the one-year-olds, will produce a higher flatter yield-effort curve that allows around 80% of the USY to be obtained for $q_i v$ ranging from 1.5 to 5 times the natural mortality rate. A net that impacts only three-year-olds and older individuals protects the fishery from overexploitation (the yield-effort curve is an increasing function of v) and at high-effort levels yields over 90% of USY, although catch per unit effort is relatively low.

The data Reed (1980) used to analyze the Arcto-

TABLE 4.2. Data for the Arcto-Norwegian cod fishery[a].

i	1	2	3	4	5	6
s_i	—	0.72	0.72	0.72	0.72	0.70
w_i	—	—	0.20	0.53	1.01	1.60
b_i	—	—	—	—	—	—
i	7	8	9	10	11	12
s_i	0.60	0.50	0.50	0.61	0.68	0.61
w_i	2.35	2.91	3.58	4.22	4.82	5.37
b_i	0.10	0.38	0.94	1.76	3.05	4.63

[a] The units associated with the weight-at-age parameters w_i and the scale of the fecundity parameters b_i are not indicated in Reed (1980).

Norwegian cod fishery is listed in Table 4.2. He also used the Ricker stock-recruitment relationship (see equation (3.10)),

$$x_1 = 18.08 x_0 e^{-0.0174 x_0}.$$

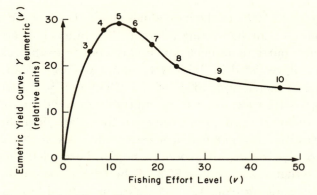

FIGURE 4.4. The knife-edge eumetric yield curve for the exploitation of Arcto-Norwegian cod. The points on the curve corresponding to integer values for t_c are indicated (adapted from Reed 1980).

The eumetric yield curve for this fishery is illustrated in Figure 4.4. Reed calculated the USY for this fishery to be $Y° = 40$ relative units, which is obtained by harvesting 45% of the six-year-olds and taking nothing from any of the other age classes. The maximum value for $Y(v, t_c)$ is 30 units, occurring at a point corresponding to removing just over 10% each year of all individuals of 5 years and older.

4.4 DETERMINISTIC DYNAMIC HARVESTING

4.4.1 Maximum Yield

Let us now return to the dynamic population harvesting model given by equations (4.9) to (4.14); that is,

$$
\begin{aligned}
x_1(k+1) &= s_0 x_0(k)\psi\big(x_0(k)\big), \\
x_{i+1}(k+1) &= e^{-(\alpha_i + q_i v(k)\hat{t})} x_i(k), \quad i = 1, \ldots, n-2, \\
x_n(k+1) &= e^{-(\alpha_{n-1} + q_{n-1} v(k)\hat{t})} x_{n-1}(k) \\
&\quad + e^{-(\alpha_n + q_n v(k)\hat{t})} x_n(k),
\end{aligned}
\tag{4.22}
$$

where $x_0(k) = \sum_{i=1}^n b_i x_i(k)$. Consider the problem of maximizing the harvest over a predetermined time period $[0, T]$. If $Y_k(v(k), \mathbf{q})$, as expressed in (4.15), is the harvest obtained during time period $[k, k+1]$, the total harvest over $[0, T]$ for given $\mathbf{x}_0$ and specified $\mathbf{x}(T)$ is

$$
J_T\big(\mathbf{x}_0, \mathbf{x}(T)\big) = \sum_{k=0}^{T-1} Y_k\big(v(k), \mathbf{q}\big).
\tag{4.23}
$$

In Chapter 3 we discussed the problem of optimizing over a finite planning horizon and concluded that, for fixed endpoint problems with zero discounting ($\delta = 1$, $L \equiv 0$ in expression (3.68)), it is appropriate to select $\mathbf{x}(T) = \mathbf{x}^*$, where $\mathbf{x}^*$ is the state of the resource corresponding to the MSY solution. We thus have the following problem:

Problem 5 (Fisheries Harvesting Problem) *Choose an appropriate sequence of harvesting effort levels* $\{v(k) \geq 0, \ k =$

$1, \ldots, T - 1\}$ *to maximize* $J_T(\mathbf{x}_0, \mathbf{x}(T))$*, expressed in (4.23), subject to system (4.22) with given initial condition* $\mathbf{x}(0) = \mathbf{x}_0$ *and specified final time condition* $\mathbf{x}(T) = \mathbf{x}^*$.

4.4.2   Suboptimal Policies

Before Problem 5 can be solved, we need to ensure that it does indeed have a solution. Whenever a problem is formulated with given initial ($\mathbf{x}_0$) and final time ($\mathbf{x}(T)$) conditions, we need to ensure that at least one sequence of controls $\{v(k) \geq 0, \ k = 1, \ldots, T - 1\}$ exists that can drive the system from $\mathbf{x}_0$ to $\mathbf{x}(T)$. This is particularly critical for the Fisheries Harvesting Problem because of the difficulty in manipulating the structure of the population vector $\mathbf{x}(k)$ using a scalar control $v(k)$, as modeled in equations (4.9) and (4.12) (the system with full vector control is modeled by equations (4.21)). For the dynamic yield problem associated with expression (3.62) we proved, under certain conditions (Theorem 3.3), that the optimal solution involves driving the population to MSY in one time step. This is no longer possible. In fact, the optimal solution will invariably involve a different value for $v(k)$ in each time step.

There are a number of undesirable side effects associated with managing a fishery by regulating the level of $v(k)$ in each time step to be a certain value. Effort level translates into units such as boat days, and if $v(k)$ is larger than the present capacity of the fishing fleet, then capitalization may occur at time k, only to lie idle in subsequent time periods $k_i > k$ for which $v(k_i) < v(k)$ (see Clark, 1985, for a discussion on capitalization). To avoid this problem, we can consider the problem of finding the constant fishing effort level v that maximizes yield for the entire time period $[0, T]$. The utility of such a *suboptimal* approach can be fully evaluated only if the yield under this suboptimal policy can be compared with the maximum yield solution over $[0, T]$. An even simpler approach is to select $v = v^*$, the solution to the MSY problem, and compare it to the maximum yield solution over $[0, T]$. Of course, one also has to take into account the fact that v^* will not (in general) drive that system to the specified $\mathbf{x}(T)$.

Other suboptimal policies exist. Instead of trying to drive that population to the MSY state vector $\mathbf{x}^*$ at time T, it would be much less stringent to drive only the scalar stock variable x_0 to the corresponding MSY stock level $x_0(T) = x_0^*$. In fact, if we do this as rapidly as possible, once $x_0(k) = x_0^*$ is obtained, recruitment in the following period will be at the MSY level. If we can maintain this for several time periods, the overall population structure should begin to approach $\mathbf{x}^*$. A policy where we try to maintain the stock at a given level is termed a *fixed-escapement* or *fixed-base level* policy and is defined as follows: if $v(k)$ is constrained to satisfy $0 \leq v(k) \leq \bar{v}$ for $k = 0, \ldots, n-1$, then

$$v(k) = \begin{cases} \bar{v}, & \text{if } x_0(k) \geq x_0^* \text{ and } x_0(k+1) \geq x_0^*; \\ 0, & \text{if } x_0(k) \leq x_0^* \text{ and } x_0(k+1) \leq x_0^*, \end{cases}$$

otherwise $v(k) \in (0, \bar{v})$ and it has the value which gives $x_0(k+1) = x_0^*$. Under stochastic conditions a fixed-escapement policy has the desirable property, as will be discussed in the next section, of providing the stock with a measure of protection from becoming overexploited.

Finally, if Problem 5 has no solution because it is impossible to meet the final time condition $\mathbf{x}(T) = \mathbf{x}^*$, then this final time condition can be "softened" by dropping it, but adding a penalty to expression (4.23) for not satisfying this final time condition. One possible penalty is based on minimizing the weighted-sum-of-squares deviation of $\mathbf{x}(T)$ from $\mathbf{x}^*$; that is,

$$J_T(\mathbf{x}_0, \mathbf{x}(T)) = \sum_{k=0}^{T-1} Y_k(v(k), \mathbf{q}) - \sum_{i=1}^{n} d_i(x_i(T) - x_i^*)^2, \quad (4.24)$$

TABLE 4.3. Maximum sustained yield solution for the anchovy fishery characterized in Table 4.1.

Weight at age	$\mathbf{w} = (1.6, 6.9, 14, 19, 24)'$ (g)
Fishing effort	$v^* = 1.15$
Yield	$Y^* = 122$ (tmt)[a]
State[b]	$\mathbf{x}^* = (79.4, 29.7, 10.1, 3.29, 0.87)'$
Spawning stock index	$x_0^* = 205$ (tmt)[a]

[a] Thousand metric tons.

[b] Each age-class has the units 10^9 individuals.

where suitable values are assigned to the penalty constants d_i.

4.4.3 Anchovy Fishery

We return to the anchovy fishery characterized by the parameters in Table 4.1, except that the growth function $w(t)$ is replaced by a constant weight at age vector $\mathbf{w}$ listed in Table 4.3. For this fishery, the MSY solution $(\mathbf{x}^*, v^*)$ is found to have the values listed in Table 4.3 (also see Figure 4.3).

Fixed endpoint problems like Problem 5 are difficult to solve numerically because we can search only among those sequences of controls $\{v(k) \geq 0, \ k = 1, \ldots, T - 1\}$ that drive the system from $\mathbf{x}_0$ to $\mathbf{x}^*$ at time T. One approach is the penalty function method which uses criterion (4.24), rather than (4.23), and thus does not explicitly ensure that the final time condition $\mathbf{x}_T = \mathbf{x}^*$ is met. If it is possible to meet this condition and the penalty constants d_i are appropriately chosen, then the boundary condition can be met to a desired level of accuracy without the penalty term contributing more than a fraction of a percent to the total value of J_T.

For the fishery considered here, it was only possible to make the penalty term insignificantly small for $T \geq 5$. This makes intuitive sense if we consider that we are trying to hit a five-dimensional target point and would expect to be able to achieve this only with at least five degrees of freedom in choosing the appropriate control (it also takes five years for the first batch of recruits to enter the oldest age class as set by the initial conditions). A gradient technique (related to Pontryagin's Maximum Principle) was used to solve the unconstrained numerical optimization problem associated with criterion (4.24) (as discussed in Getz, 1985; also see Chapter 5), and a good approximate solution was obtained for $d_i = 10^{-11}$, $i = 1, \ldots, 5$. With these constants the boundary condition matched x^* as listed in Table 4.3 so that in penalty contributed less than 0.2% of the yield when $T = 5$, and 0.01% of the yield when $T = 10$.

The five- and ten-year optimal policies were compared with the performance of the fishery under constant-MSY-effort and fixed-MSY-escapement policies for two cases. In the first case the initial condition was taken to be the MSY level x^*, except for $x_1(0)$ which was set to the maximum recruitment level as determined by the Beverton and Holt model parameters in Table 4.1; that is, $x(0) = (122, 29.7, 10.1, 3.29, 0.87)'$ (see Table 4.3). In the second case the initial condition was taken to be the MSY level x^*, except for $x_1(0)$ which was set to the minimum possible recruitment level; that is, $x(0) = (0.0, 29.7, 10.1, 3.29, 0.87)'$ (see Table 4.3). The results obtained for these two cases are listed in Tables 4.4 and 4.5, respectively. Note that endpoint conditions are not specified for the constant-MSY-effort and fixed-MSY-escapement policies, although the state $x(t)$ will approach x^* if the system is stable when harvested using these policies.

The most striking feature of these results is that after ten years, the constant-MSY-effort and fixed-MSY-escapement policies performed almost as well the optimal ten-year policy in both cases. The fact that the constant-effort policy outperformed the optimal five-year policy in the failed recruitment case (Table 4.5) is due to the fact

159

TABLE 4.4. Yields[a] and corresponding effort levels in parenthesis for the maximum recruitment case $\mathbf{x}(0) = (122, 29.7, 10.1, 3.29, 0.87)'$.

Year	Constant effort	Fixed escapement	10-year opt.	5-year opt.
1	131(1.15)	157(1.43)	117(1.01)	118(1.03)
2	143(1.15)	189(1.76)	178(1.44)	169(1.36)
3	141(1.15)	120(1.19)	154(1.35)	146(1.24)
4	141(1.15)	118(1.09)	140(1.25)	155(1.36)
5	131(1.15)	124(1.17)	123(1.17)	121(1.15)
Average[b]	134.7(1.15)	141.5(1.33)		141.8(1.23)[c]
6	128(1.15)	122(1.15)	121(1.14)	
7	126(1.15)	122(1.15)	120(1.13)	
8	125(1.15)	122(1.15)	120(1.13)	
9	124(1.15)	122(1.15)	126(1.18)	
10	123(1.15)	122(1.15)	121(1.18)	
Average[d]	131.3(1.15)	131.7(1.24)	131.9(1.20)[e]	

[a] Thousand metric tons.

[b] Taken over first 5 years.

[c] Penalty<0.01 of yield.

[d] Taken over all 10 years.

[e] Penalty=0.18 of yield.

that after five years the constant effort policy has not yet allowed the stock to recover to MSY levels; that is, under the MSY policy the stock remains in a weakened state for a longer period of time. Also notice that when the stock is strong the optimal ten-year policy requires marginally more effort than the constant-effort policy to achieve maximum yield, while the ten-year policy requires marginally less effort than the constant-effort MSY when the stock is weak.

Another striking feature of the results is how markedly different the three policies are from year to year, even though their long-term performance is so similar (see Figure 4.5). Thus if the stock was truly stochastic so that harvesting policies needed to continually respond as strong- and weak-year classes passed through the fishery, constant-effort and constant-escapement policies would differ sig-

TABLE 4.5. Yields[a] and corresponding effort levels in parentheses for the maximum recruitment case $\mathbf{x}(0) = (0, 29.7, 10.1, 3.29, 0.87)'$.

Year	Constant effort	Fixed escapement	10-year opt.	5-year opt.
1	105(1.15)	63(0.63)	115(1.28)	115(1.29)
2	82(1.15)	0(0.00)	26(0.33)	33(0.42)
3	85(1.15)	93(0.78)	65(0.67)	58(0.60)
4	82(1.15)	136(1.37)	82(0.87)	54(0.54)
5	99(1.15)	116(1.09)	118(1.08)	157(1.40)
Average[b]	90.6(1.15)	81.5(0.77)		83.4(0.85)[c]
6	104(1.15)	123(1.15)	125(1.14)	
7	109(1.15)	122(1.15)	124(1.15)	
8	113(1.15)	122(1.15)	123(1.14)	
9	116(1.15)	122(1.15)	128(1.20)	
10	118(1.15)	122(1.15)	121(1.15)	
Average[d]	101.4(1.15)	101.8(0.96)	102.7(1.00)[e]	

[a] Thousand metric tons.

[b] Taken over first 5 years.

[c] Penalty<0.01 of yield.

[d] Taken over all 10 years.

[e] Penalty=0.03 of yield.

nificantly over the short term. (This question is pursued further in the next section.)

Here we have analyzed the performance of the optimal and suboptimal policies only with respect to yield. We could evaluate the performance of these policies with respect to other criteria such as maximizing rent (gross value of yield minus operating costs), or the present value of the fishery (sum of all discounted future rents). In the latter case the optimal policy could depend on price, cost, and discount parameter values (Clark, 1985) and, of course, it would also depend on the length of the planning horizon (see Chapter 3).

4.4.4   Rehabilitation of Fisheries

Uncontrolled fisheries often lead to overexploited stocks. For example, Atlantic menhaden (*Brevoortia tyrannus*) and Pacific Ocean perch (*Sebastes alutus*) stocks have

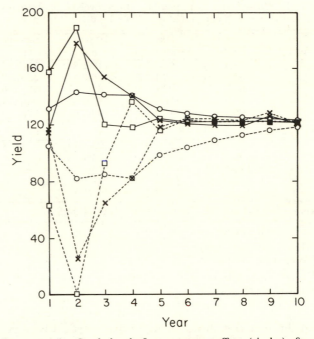

FIGURE 4.5. Catch levels from constant-effort (circles), fixed-escapement (squares), and ten-year optimal (crosses) harvesting policies for cases A (initial maximum recruitment, solid lines) and B (initial zero recruitment, broken lines).

been overfished for several years and reductions in fishing have been recommended (Hightower and Grossman, 1987). A stock rehabilitation program is essentially an infinite time horizon optimal-harvest problem, although a number of different finite time horizon *ad hoc*, but reasonable, planning approaches can be taken. For example, Huang and Walters (1983) used an age-structured model to compare the performance of constant-effort rehabilitation policies versus short-term (3-year) closure of the large yellow croaker (*Pseudosciaena crocea*) fishery in the China Sea, while Archibald *et al.* (1983) compared several constant-effort policies for the rehabilitation of Pacific Ocean perch.

Hightower and Grossman (1987) took a more elaborate approach to the problem by finding the optimal rehabilitation schedules for several formulations, each emphasizing a different aspect of the management problem. The model they used is essentially equations (4.22), where $\psi(x_0)$ was selected to have either the Beverton and Holt or Ricker (equations (3.9) and (3.10)) form for the stock-recruitment relationship. In Section 3.4 we discussed at some length the problem of choosing an appropriate final time condition when planning to manage a resource over a finite planning horizon. Hightower and Grossman (1987) decided in their analysis that the stock should be driven towards the state $\mathbf{x}^{0.1}$ corresponding to $v^{0.1}$ (the point on the yield-effort curve—as explained in Section 4.3—at which the slope is one-tenth of the slope at the origin).

Besides considering the *maximum-yield problem* of maximizing criterion (4.24), with x_i^* replaced by $x_i^{0.1}$, subject to equations (3.50) and a given initial condition $\mathbf{x}_0$, they considered a *maximum-log-yield problem* (maximizing the logarithm of (4.24); also see Ruppert *et al.*, 1985) with a penalty associated with not meeting the final time condition, and minimization problems associated with the following two criteria:

Minimum-stock-deviation criterion

$$J_T(\mathbf{x}_0, \mathbf{x}_T) = \sum_{k=0}^{T-1} (x_0 - x_0^{0.1})^2. \qquad (4.25)$$

Minimum-yield-deviation criterion

$$J_T(\mathbf{x}_0, \mathbf{x}(T)) = \sum_{k=0}^{T-1} \left(Y_k(v(k), \mathbf{q}) - Y^{0.1} \right)^2$$

163

$$+ \sum_{i=1}^{n} d\left(x_i(T) - x_i^{0.1}\right)^2, \qquad (4.26)$$

where $Y^{0.1}$ is the equilibrium corresponding to $v^{0.1}$, and d is a penalty constant. Note that they actually solved two problems associated with criterion (4.25). In one case they interpreted the variable x_0 as the biomass of the total stock, and in the other case they interpreted x_0, as we have, as the biomass of the spawning stock.

Hightower and Grossman (1987) found optimal stock rehabilitation policies, for three species mentioned above (for the anchovy parameters in Table 4.1 above, and for the menhaden and perch parameters in Tables 4.6 and 4.8 in Section 4.5 below), using criteria (4.24), (4.25) and (4.26). They found that heavily exploited stocks could be rehabilitated within two to three times the life span of parameter n. Rehabilitation periods, however, were longer for anchovy ($n = 5$) than for menhaden ($n = 8$). Hightower and Grossman concluded that this was probably due to the fact that maximum recruitment rates for menhaden, as determined by a dome-shaped Ricker stock-recruitment curve, are reached rapidly while maximum recruitment for anchovy, as determined by an asymptotically increasing Beverton and Holt stock-recruitment curve, is never reached. Pacific Ocean perch are long lived ($n = 29$) and needed 60 years without fishing to recover to $\mathbf{x}^{0.1}$ stock levels.

The optimal minimum-stock-deviation policies are qualitatively the same for all three stocks, that is, to close the fishery until the stock target levels are reached. The minimum-yield-deviation policies resulted in a more gradual recovery for all three stocks. If the fisheries were closed at the start of the planning horizon, however, the optimal minimum-yield-deviation policies allowed $\tilde{v}(k)$ (the optimal effort sequence) to exceed $v^{0.1}$ once the fishery reopened. Relatively high effort levels may be undesirable if they lead to capitalization of the fishery to the point where a number of boats lie idle in years when the stock is low. This problem can be avoided by constraining the maximum value of $v(k)$. Hightower and Grossman found that policies that maximized harvest appeared to have certain desirable qualities not met in certain cases by the other policies. In case of the heavily depleted menhaden and perch stocks, for example, these policies were the only ones that did not require closure of the fishery during some part of the planning horizon.

Hightower and Grossman (1987) also introduced random variation into the recruitment process, but they were no longer able to solve for optimal rehabilitation policies. Using Monte Carlo simulation techniques (as discussed in Section 3.5), however, they concluded that deterministic harvest policies that responded adaptively to stochastic changes in the stock (the effort level is recalculated each time the current state of the stock is measured) could perform very well provided that the updating occurred frequently enough (also see Walters, 1986). Thus deterministic policies have a place in fisheries if stock levels are regularly monitored.

4.5 STOCHASTIC HARVESTING

4.5.1 A Stochastic Cohort Model

The cohort model presented in Section 4.2 assumes that recruitment $x_1(k)$ is constant independent of the stock index $x_0(k-1)$ at time of spawning. If the environment or spawning habitat is limiting, as could well be the case if the stock is relatively large, then recruitment can be modeled as a constant R that is modified by a stochastic variable Z_k; the latter reflecting the stochastic changes in the environment that occur in the $(k+1)$-th year. The variables

165

Z_k are assumed to be *i.i.d.* with unit mean. As will be discussed later, this approach may lead to an overestimate of the long-term productivity of the fishery.

More generally, the survival of age class i from one time period to the next may also be stochastic; that is, the parameter s_i is replaced by stochastic variables S_{ik} such that for $i = 1, \ldots, n$, the variables $\{S_{ik} | k = 0, 1, \ldots\}$ are assumed to be *i.i.d.* with mean s_i. (Note that if identity (4.13) is assumed to hold, then α_i could be regarded as stochastic.) For the case where we assume that each age class can be separately harvested, Mendelssohn (1978) obtained the following result.

Result 4.1 *Consider the problem of harvesting a population modeled by*

$$X_1(k+1) = Z_k R(k) - u_1(k)$$
$$X_{i+1}(k+1) = S_{ik} X_i(k) - u_{i+1}(k), \qquad i = 1, \ldots, n,$$

where Z_k and S_{ik} are stochastic variables as defined above. Then the problem

$$\max_{\mathbf{u}(0), \cdots, \mathbf{u}(T-1)} E\left[\sum_{t=0}^{T-1} \mathbf{w}' \mathbf{u}(k)\right]$$

can be considered as n separate cohort harvesting problems, where the solution to harvesting individuals in the cohort born at time k is to remove all individuals in that cohort at the age at which individuals in the cohort achieve their maximum expected biomass (this solution is directly analogous to the deterministic cohort problem). If this occurs at age $\hat{i}$, then $u_j(k) = 0$, for all $j \neq \hat{i}$ and $k \leq T - \hat{i}$. If the latter inequality is not satisfied, then the particular cohort is born too close to the end of the planning horizon to achieve its maximum expected biomass over the period $[k - j, T]$. Such cohorts are harvested at the point where they achieve their maximum biomass over this reduced period, which typically will occur at time T.

166

Mendelssohn (1978) also considered the problem where each $u_i(k)$ cannot be uniquely determined but is related to a scalar effort variable $v(k) \in [0, \bar{v}]$ by the equation

$$u_i(k) = v(k)q_i(k)x_i(k). \tag{4.27}$$

He proved that in the kth time interval the optimal policy is either $v(k) = 0$ or $v(k) = \bar{v}$; that is, the optimal solution corresponds to *pulse fishing* (the optimal solution is a sequence of maximum and zero effort levels).

Mendelssohn (1978) obtained these results using dynamic programming (see the Principle of Optimality in Chapter 3) and used the same approach to identify conditions when it is always optimal to first harvest the oldest individuals in the population. His approach, however, does not deal with situations in which cohorts interact with each other as they would if a stock-recruitment relationship existed. Except for the linear case, such situations are exceedingly difficult to analyze, although Lovejoy (1986) managed to obtain bounds on the optimal age-at-first-capture for models that include nonlinear density-dependent recruitment. In most cases, the analysis of such stochastic nonlinear systems is carried out for specific problems using numerical simulation techniques.

4.5.2 A Nonlinear Stochastic Recruitment Model

In stochastic fisheries without age structure (lumped), Reed (1979) has shown that a fixed-escapement policy maximizes yield. Analyses of stock and yield variance have been carried out using nonlinear lumped models (Beddington and May, 1977, May *et al.*, 1978), as well as age-structured models with stochastic recruitment for systems linearized around their equilibrium (Horwood, 1982, 1983). The latter involves Fourier analysis techniques and will not be covered here. Cohen (1976, 1977a, 1977b, 1979) has extensively analyzed the dynamic properties of stochastic Leslie matrix models. He and his colleagues have analytically computed long-term measures of the growth rate for striped bass (*Morone saxatilis*) using a simplified stochastic Leslie model that includes an egg index and reproductive adults (that is, two age classes) (Cohen *et al.*, 1983). The problem with this density-dependent approach, of course, is that long-term growth is not limited by environmental considerations and recruitment is unbounded.

In general, numerical simulations can be used to obtain variance estimates, but numerical solutions to optimal stochastic age-structured harvesting problems are often intractable. The performance of various suboptimal policies, however, are easily evaluated numerically. This approach has the drawback that we do not know how well our suboptimal policies perform with respect to the optimal policy but, at least in the deterministic case, we have shown above that suboptimal policies can be within 99% of the optimal yield. Furthermore, the critical question is not whether we can squeeze another 1% or 2% yield out of the resource, but what the effects of variability are on the stability and operation of the fishery.

There are basically two approaches to the problem of numerically simulating the dynamics of harvesting a stochastic stock. The first is to simulate the means and variance/covariance terms of the stock and yield using a model such as system (3.93). This approach was taken by Getz and Swartzman (1981) and Swartzman *et al.* (1983) in their analysis of the management of Pacific whiting (*Merluccius productus*). The model they used, however, was a stochastic transition matrix approach that provided a cruder approximation to the underlying stochastic process than is provided by system (3.93). This system of equations has been used to analyze the whiting fishery under a U.S./Canadian "two player" exploitation. This analysis is presented in the final section below.

The second approach is to use Monte Carlo simulation techniques to derive estimates of the means and variance/covariance terms of the stock and yield associated with the exploitation of a particular fishery. This approach has the advantage of not requiring the derivation of a system of equations such as (3.93) for the moments of the process modeled by equations (3.87). The disadvantage, however, is the amount of computer time required for the study, especially if numerical optimization techniques are involved. The Monte Carlo approach has been used in a number of studies (Hightower and Grossman, 1985; Ruppert *et al.*, 1985; Getz, Francis, and Swartzman, 1987; Hightower and Lenarz, in press). The results of these studies are summarized here.

Almost invariably, analyses of age-structured stochastic fisheries have assumed a deterministic transition model with stochastic recruitment, that is, a model such as system (3.87), where natural (s_i) and fishing $(1 - h_i)$ mortality proportions are of the form expressed in (4.13) and (4.14). Specifically, the model is given by

$$X_0(k) = \sum_{i=1}^{n} b_i X_i(k)$$

$$X_1(k+1) = Z_k \phi\big(X_0(k)\big)$$

$$X_{i+1}(k+1) = e^{-(\alpha_i + q_i v_k)} X_i(k), \qquad i = 1, \ldots, n-2,$$

$$X_n(k+1) = e^{-(\alpha_{n-1} + q_{n-1} v_k)} X_{n-1}(k)$$

$$+ e^{-(\alpha_n + q_n v_k)} X_n(k), \qquad (4.28)$$

where ϕ is the total recruitment function (linear recruitment function multiplied by the nonlinear modifier function ψ; cf. discussion above equation (3.15)) and Z_k are the *i.i.d.* stochastic variables.

4.5.3 Constant Harvesting Policies

Hightower and Grossman (1985) considered the problem of harvesting South African anchovy (see Tables 4.1

and 4.3), Atlantic menhaden, and Pacific Ocean perch stocks at different constant levels of fishing effort. They assumed that

$$Z_k = e^{\nu_k}, \tag{4.29}$$

where ν_k are drawn from the normal distribution $N(0, \sigma^2)$ (zero mean, variance σ^2). They justify this choice of distribution because the term "... e^{ν_k} can be viewed as a random survival factor resulting from several independent and multiplicative environmental factors (Walters and Hilborn, 1976). Thus, ν represents the sum of several random factors and should be approximately normally distributed by the Central Limit Theorem (Walters and Hilborn, 1976)." The lognormal distribution also has the biologically required property of nonnegativity and is able to produce an occasional year class that is considerably larger than the mean, as is observed in empirical data sets (see Hennemuth *et al.*, 1980; Peterman, 1981).

Hightower and Grossman (1985) obtained estimates of σ^2 for thirteen marine fish stocks by regressing the log-transformed equation (so that ν_k appeared linearly in the equations) of the constant, Beverton and Holt, and Ricker forms of the stock-recruitment function ϕ, using available data. On average, 23 years of data (range 13 to 46) were available for each of the thirteen sets of data, and estimates of σ^2 for the Beverton and Holt and the Ricker models ranged from 0.06 to 1.20. Most values for σ^2 were below 0.75 so that Hightower and Grossman selected $\sigma^2 = 0.25, 0.50, 0.75$ to respectively represent low, moderate, and high levels of variability to simulate the effects of variability on constant-effort management policies.

For the anchovy fishery, Hightower and Grossman (1985) used the same model parameter set listed in Table 4.1. In addition they indicate an observed lognormal

TABLE 4.6. Parameter values for the Atlantic menhaden fishery.

Number of age-classes	$n = 8$
Length of harvesting season	$\bar{t} = 1$
Natural mortality parameters	$\alpha_i = 0.25, i = 1, \ldots, 8$
Weight-at-age[a]	103, 260, 412, 530,
	614, 671, 707, 731
Fecundity values[b]	0, 110, 227, 303,
	355, 410, 492, 497
Stock-recruitment function[c]	$x_1 = e^\mu 0.0205 x_0 e^{-0.0024 x_0}$
Relative catchability coefficients[d]	0.57, 1.81, 1.58, 1.54,
	1.43, 1.82, 1.70, 1.70

[a] w_i listed by age $i = 1, \ldots, 8$; units are grams.

[b] b_i listed by age $i = 1, \ldots, 8$; units are grams, that is, x_0 is in grams.

[c] The unit is numbers of individuals. The lognormal variance associated with this function is estimated to be $\sigma^2 = 0.25$.

[d] q_i listed by age $i = 1, \ldots, 8$; see Hightower and Grossman (1985) for more details.

variance of $\sigma^2 = 0.06$ associated with the Beverton and Holt stock-recruitment function. The Atlantic menhaden parameter set that Hightower and Grossman used is listed in Table 4.6. Since Pacific Ocean perch is a relatively long-lived species (Hightower and Grossman chose $n = 29$), the full set of parameters is not listed here. A range of weight-at-age and the estimated natural mortality rates for this species are listed in Table 4.8, Section 4.5.5, below. Hightower and Grossman used a different stock-recruitment function to that associated with the parameters in Table 4.6. Specifically, they used the stock-recruitment function $x_1 = e^\mu 0.0883 x_0 e^{-9.01 \times 10^{-5} x_0}$, and associated lognormal variance $\sigma^2 = 0.30$, which they obtained from estimates made by Archibald *et al.* (1983).

Although Hightower and Grossman (1985) have estimates of the variance associated with the stock-recruitment functions for the three fisheries in question, for purposes of comparison they used low, moderate, and high levels of variability, as defined above. For each fish-

171

ery, they also found the effort level $\hat{v}$ that gave the maximum expected sustainable yield (MESY[1]) solution and, for comparative purposes, evaluated the response of the three fisheries to 1.25, 1.5, 1.75, and 2.0 times the corresponding $\hat{v}$. To obtain MESYs for the three fisheries, 100 replicates for each level of constant-effort (over an appropriately chosen range of effort levels; see Hightower and Grossman, 1985, for details) was run for $10n$ years (n respectively equals 5, 8, 29 for the anchovy, menhaden, and perch fisheries); but the average annual yield was calculated only over the last $5n$ years to avoid the influences of the nominal set of initial conditions used in each simulation.

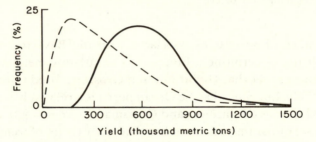

FIGURE 4.6. Frequency distributions of harvest for Atlantic menhaden at moderate levels of environmental variability ($\sigma^2 = 0.50$) for two constant levels of fishing effort: $\hat{v}$, solid line; $2\hat{v}$, dotted line. (Smoothed representation of data adapted from Hightower and Grossman, 1985.)

The results indicate that increasing environmental variability tends to obscure differences between policies with

[1] Throughout the text we refer to MESY as the solution that corresponds to the expected maximum yield under constant-effort policies. There are other classes of policies that satisfy a time-averaged equilibrium condition. To avoid confusion, we will not refer to those that maximize the yield in a particular class of time-averaged equilibrium yield policies as MESY policies, since the maximum with respect to each class may be different.

different levels of constant fishing effort. But even at low levels of environmental variability, these differences are evident only when v is at least 50% larger than $\hat{v}$ (Figure 4.6). Thus the maximum yield that can be obtained is robust to moderate increases in effort level above $\hat{v}$, and any decreases are increasingly difficult to detect with greater levels of stochasticity. As a consequence of this, stochasticity in recruitment provides the manager with some leeway in incorporating the demands of fishermen and other socioeconomic factors into a plan for regulating a fishery. Note, however, that the drop in yield becomes apparent even at high levels of stochasticity ($\sigma^2 = 0.75$, say) if v is increased beyond $1.5\hat{v}$.

4.5.4 Escapement-Related Policies

In contrast to a constant-effort policy, there is the fixed-escapement policy discussed above in the context of harvesting deterministic dynamic stocks (see Figure 4.5). In order to implement such a policy, we need to assume that we can monitor fishing so that a fixed number of fish are left unharvested at the end of the fishing season. Depending on the type of fishery, a direct assessment of escapement may be possible, as in the case of counting salmon heading upstream to spawn. More generally, escapement levels can be inferred by changes in catch per unit effort throughout the fishing season and/or estimates of the stock prior to the beginning of the season and measurement of the catch as the season progresses. However, these techniques for predicting escapement levels are likely to be very inaccurate.

Assuming that it is possible to set yield levels (catch quotas) by monitoring the stock and ensuring that a given stock level (biomass or egg index; see term in parentheses in equation (4.17)) is achieved, then a number of policies, other than fixed-escapement, could be based on escapement levels. For example, Ruppert *et al.* (1985) proposed

173

a policy where the catch is determined by the yield equation

$$Y_k = \begin{cases} \tilde{q}(x_0(k) - \tilde{x}_0)^\rho, & \text{if } x_0(k) \geq \tilde{x}_0; \\ 0, & \text{otherwise,} \end{cases} \qquad (4.30)$$

where $\tilde{q}$, $\tilde{x}_0$, and ρ are appropriately chosen constants. Note that equation (4.30) is quite general, since it includes constant-catch ($\rho = 0$), fixed-escapement ($\rho = 1$, $q = 1$, and x_0 is a total stock rather than a spawning stock variable), and constant-effort policies ($\rho = 1$, $\tilde{x}_0 = 0$, and assuming that age-specific catchability is proportional to qb_i and that this quantity is much smaller than 1; cf. (4.11) and (4.27)).

Ruppert *et al.* (1985) used Monte Carlo simulation techniques to determine, among other things, the values of $\tilde{x}_0$ and q that maximized long-run average yield and also the long-run average of the logarithm of the yield (log-yield criterion obtained by taking the sum of $\log(Y_k)$ instead of Y_k in expression (4.23)) for the two cases $\rho = 1$ (fixed-escapement policy) and $\rho = 0.5$ (a compromise between constant-catch and fixed-escapement). Details of the numerical algorithm that they used are given in Ruppert *et al.* (1984). The log-yield criterion penalizes very small yields and leads to a so-called *risk averse* or low variance yield solution (see Ruppert *et al.*, 1984, 1985, for more details). The parameters they used did differ from those given in Table 4.6 and they used a Beverton and Holt rather than a Ricker stock-recruitment relationship. Note that with some algebraic manipulation, the Beverton and Holt relationship expressed by equations (3.6) and (3.9) can be rewritten as

$$\frac{1}{x_1(k+1)} = \frac{1}{s_0 x_0(k)} + \frac{s_0}{\beta}. \qquad (4.31)$$

Instead of inserting a multiplicative lognormally distribute stochastic variable Z_t into their model, as indicated by the second equation in system (4.28), Ruppert *et al.* (1985) inserted an additive normally distributed stochastic variable into the inverse Beverton and Holt equation (4.31); that is, they used the equation

$$\frac{1}{X_1(k+1)} = \frac{1}{s_0 X_0(k)} + \frac{s_0}{\beta} + Z_k, \qquad (4.32)$$

where Z_k itself is drawn from the normal distribution $N(0, \sigma^2)$, subject to the constraint that $X_1(k+1) \geq 0$. Note that they linearly regressed equation (4.32) onto the appropriately transformed stock-recruitment data set to obtain an estimate of σ^2 for their simulation (see Ruppert *et al.*, 1985, for further details).

In this particular study, they made 250 Monte Carlo runs to estimate the average yield and standard deviation that corresponded to a particular harvest policy. The yield results for the best constant-effort, approximate fixed-escapement, and compromise fixed-escapement/constant-catch policies are listed in Table 4.7. From this table it is apparent that the escapement-based policies provide at least 15% more yield than the best constant-effort or MESY policy. It is possible that this result is in error, as Hightower and Lenarz (in press) were not able to verify this result using a similar approach. Bergh and Butterworth's (1987) stochastic management analysis of the South African anchovy fishery also showed very little difference between constant-effort and fixed-escapement policies. Also the maximum yield criterion should always provide a larger average yield than the yield associated with the log-yield criterion, but this is not the case for the compromise fixed-escapement/constant-effort polices in Table 4.7 (the numerical algorithm employed may not be precise enough to accurately determine the third significant digit). Note that the yield maximizing fixed-escapement policy has almost twice the standard deviation

175

TABLE 4.7. Yield results for the best policies for harvesting Atlantic menhaden. After Ruppert *et al.* (1985).

Policy	Optimal parameters[a]	Yield[b]	Standard deviation[b]
Constant-effort	$\hat{v} = 0.8$	450	220
	Yield criterion		
Fixed-escapement ($\rho = 1.0$)	$\bar{q} = 2.36$ $\bar{x}_0 = 77.6$	528	406
Compromise fixed-escapement and constant-catch ($\rho = 0.5$)	$\bar{q} = 12.0$ $\bar{x}_0 = 124$	525	221
	Log-yield criterion		
Fixed-escapement ($\rho = 1.0$)	$\bar{q} = 1.21$ $\bar{x}_0 = -80.5$	527	244
Compromise fixed-escapement and constant catch ($\rho = 0.5$)	$\bar{q} = 11.0$ $\bar{x}_0 = 108$	527	203

[a] See equation (4.30).

[b] Thousand metric tons.

in yield when compared with any of the other four policies. This level of variation can be reduced, as mentioned before, by selecting one of the more conservative policies associated with $\rho = 0.5$ in expression (4.30), or by maximizing the log-yield rather than yield criterion.

A more general approach to setting yield levels has been proposed by R. C. Myers (unpublished manuscript). He suggests including an autocorrelation parameter in generating the stochastic effects of environmental noise and a scheme for updating estimates of the parameters associated with the stock-recruitment processes. He also proposes that the three-parameter family of yield-determining curves expressed in equation (4.30) be replaced with the following four-parameter family:

$$Y_k = \left(\frac{\tilde{q}_1 \left(x_0(k) - \tilde{x}_0 \right)}{\tilde{q}_2 + \left(x_0(k) - \tilde{x}_0 \right)} \right)^{\rho}.$$

Furthermore, instead of maximizing the sum of the yields or log yields, Myer proposes maximizing the sum of a a concave *yield utility* function $U(Y_k)$, such as

$$U(Y) = \frac{Y}{1 + \kappa Y},$$

where κ is an appropriately chosen constant. Of course other functional forms can be selected for Y_k and its utility. Which form is most appropriate will emerge only with experience and may depend on the particular problem at hand.

4.5.5   Allocation of Variability

In a stochastic fishery with highly variable recruitment, the stock can exhibit a large degree of variability as sequences of strong and weak year classes pass through the fishery. In Pacific whiting, for example, recruitment is strongly linked to the mean January–March sea surface temperature. The mean recruitment level in years for which the temperature is below 15.6°C (cold years) is less than 40% of the mean recruitment level for which the temperature is at or above this point (warm years). Over the period 1931–1980, at least 60% of the years were cold with strings of three or more consecutive cold years occurring during the periods 1932–1937, 1946–1956, and 1974–1976. If we shift these periods a couple of years forward to account for recruitment delays, then these periods correspond to relatively low observed stock levels, while one period of warm years (1957–1961) corresponds to a substantially strengthened stock through the mid-sixties (see Swartzman *et al.*, 1983, for details).

In the previous section we saw that escapement-type policies, which respond to stock levels, can lead to improved long-term yield levels, although yield variability may be increased. Thus it is important to respond actively when sequences of relatively strong or weak year classes pass through the fishery. It is not always economically desirable, however, to transfer the variability in the stock into the yield. On the other hand, there is substantial risk involved in continuing to exploit a stock when it is relatively weak. A number of highly productive fisheries, including Peruvian anchoveta and Californian sardine, have collapsed in the past, and it is essential to select policies that minimize the risk of this happening while still maintaining a reasonable long-term average-yield level.

These questions were addressed by Getz, Francis, and Swartzman (1987), where the performance of three types of harvest policy (constant-effort, fixed-escapement, and a policy that maximized the effort level that would be allowed under a combined constant-effort/fixed-escapement policy) was evaluated in three dynamically contrasting fisheries: the fishery on South African anchovy (see Table 4.1), a short-lived, relatively fast-growing species; the fishery on the U.S./Canada West Coast for Pacific Ocean perch, a long-lived, slow-growing species; and the fishery on U.S./Bering Sea walleye pollock (*Theragra chalcogramma*), a species that is intermediate in growth and survival with respect to the other two species. The equations used to model these stocks are (4.28) and (4.29). For purposes of comparison, it was decided that the stock-recruitment function $\phi(x_0)$ should have the same form for all three stocks and the power function (3.12) was chosen for this purpose. Since it is difficult to obtain a good estimate of the power function *shape parameter* β (the power to which the function is raised) for most fisheries, Getz, Francis, and Swartzman (1987) also investigated the sensitivity of the yield to the value of β. They also expressed yield and stock levels in units relative to the expected virgin stock biomass, x_1^0 (that is, the expected equilibrium value of x_1 for an unharvested population), thereby avoiding the necessity of estimating this parameter from catch data. If $x_r^\star$ is the expected virgin recruitment level and recruitment is related to the stock level r years ago (rather than the one-year delay used throughout our treatment), then the stock-recruitment function they used has the form (cf. the power function (3.12)),

$$x_r(k + r) = x_r^\star \big(x_0(k)/x_0^\star\big)^\rho, \qquad 0 < \rho < 1. \qquad (4.33)$$

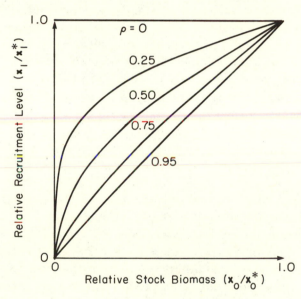

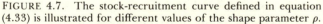

FIGURE 4.7. The stock-recruitment curve defined in equation (4.33) is illustrated for different values of the shape parameter ρ.

The shape of this stock-recruitment relationship as a function of the parameter $\rho \in [0, 1]$ is illustrated in Figure 4.7. Getz, Francis, and Swartzman (1987) investigated the sensitivity of the MSY solution to the value of ρ by carrying out deterministic equilibrium yield analyses, as discussed in Section 4.3, for all three fisheries and the values of ρ illustrated in Figure 4.7. Note that the constant recruitment case $\rho = 0$ is equivalent to Beverton and Holt's (1957) yield-per-recruit analysis. Also the linear case $\rho = 1$ is equivalent to harvesting a Leslie (1945) matrix model; but, being linear, has no nontrivial equilibrium (a nonzero yield level cannot be sustained for any effort values). Thus the sustainable yield and corresponding stock biomass curves illustrated in Figure 4.8 only apply for $\rho < 1.0$. For most stocks one would expect $\rho < 0.5$ (Kimura et al., 1984), but it is clear from Figure 4.7 that, even for these values of ρ, MSY as a percentage of the unexploited biomass level $x_0^\star$ is quite sensitive to the actual value of ρ. As ρ ranges between 0.0 and 0.5, MSY/$x_0^\star$ ranges between 0.1 and 0.2 for anchovy, between 0.05 and 0.15 for pollock, and between 0.01 and 0.03 for Pacific Ocean perch. Thus we have a two- to threefold decrease in MSY as a proportion of $x_0^\star$ as ρ ranges over $[0.0, 0.5]$. MESY estimates will likewise be affected by our choice of ρ, but it is not clear how the relative performance of different management strategies will be affected as a function of ρ.

The parameters used by Getz, Francis, and Swartzman (1987) for the three fisheries mentioned above are summarized in Table 4.8 (also see Table 4.1). The MESY estimates were obtained for each of the three fisheries by averaging yield over a number of different 1000-year Monte Carlo simulation runs, where each run corresponded to a different constant level of effort, and effort values were steadily increased from zero until a maximum of the average yields was obtained (that is, the average yields started to decrease with increased effort). Variability was expressed in terms of the coefficient of variation associated with a particular quantity (the coefficient of variation is

TABLE 4.8. Population parameters for three fisheries.

| | Stocks | | |
Parameters	South African anchovy	Bering Sea pollock	Pacific Ocean perch
Number of age classes n	5	9	13
Age r at recruitment	1	2	6
Natural mortality rate[a] α_i	0.8	0.4	0.05
Weight at:			
age r	1.6	160	454
age $n-1+r$	24	1200	1360
Virgin recruitment[b] $x_r^\star$	132,000	936	45

[a] In all three fisheries α_i has the same value for all i.

[b] This parameter is measured in numbers per metric ton of virgin biomass $x_0^\star$ (see text for more details).

given by ($\sqrt{\text{variance}}/\text{mean}$) × 100; i.e., it is the standard deviation expressed as a percentage of the mean).

One set of results obtained by Getz, Francis, and Swartzman (1987) is illustrated in Figure 4.8. These results indicate that the MESY decreases in all fisheries as the coefficient of variation (CV) of recruitment increases from 0 to 200%. The effect is negligible for Pacific Ocean perch, but is more in evidence for short-lived species such as anchovy, where the stochastic effects of varying year class strengths have a greater impact on the overall stock biomass levels available to the fishery each year. Note that these results, when compared with the MSY results obtained as a function of the value of ρ, indicate that the amount of environmental variability is much less critical than a good estimate for ρ in assessing long-term productivity. For anchovy, the most environmentally sensitive of the three stocks, the results in Figure 4.8 indicate that the

MESY is reduced by 3% when compared with MSY for the environmental CV = 100%, but the MSY itself changes by more than 50% when ρ increases from 0.0 to 0.5 (Figure 4.7). Note that Clark (1985) considers variability to be "low" when the CV is less than 50% and to be "high" when the CV is greater than 100%.

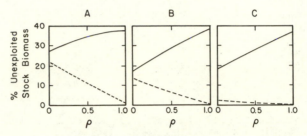

FIGURE 4.8. Maximum sustainable yield (broken line) and corresponding stock biomass levels (solid line) (as a percentage of $x_0^\star$), in terms of ρ (see equation (4.33)), for the following fisheries: A, South African anchovy; B, Bering Sea pollock; C, Pacific Ocean perch.

The three harvesting policies that were considered by Getz *et al.* (1987) are as follows: (1) $v = \hat{v}$ each year (MESY constant-effort policy); (2) v was selected so that the stock biomass at the end of the year was $\hat{x}_0$ (that is, the stock biomass corresponding to the MESY solution); or if the stock was particularly weak and this were not possible, then $v = 0$ (MESY fixed-escapement policy); (3) v was selected to be the maximum of $\hat{v}$ and the value of v that would result in a stock biomass of $\hat{x}_0$ at the end of the year (exploitative policy). The latter policy is referred to as exploitative because fishing effort is increased beyond $\hat{v}$ when the stock would otherwise be above $\hat{x}_0$, but it does not reduce v below $\hat{v}$ when the resulting stock is below $\hat{x}_0$.

These policies were evaluated by averaging over 1000-year Monte Carlo simulation runs, where the environ-

TABLE 4.9. Average yield, effort, and stock biomass levels obtained for a MESY constant-effort policy (CE), a MESY fixed-escapement (FE), and an exploitative policy (EX) using Monte Carlo simulation techniques.

Stock	Policy	Yield[a]	Effort[b]	Stock[a]	Min. effort[c]	Biomass below[d] $\frac{x_0}{2}$
Anchovy	CE	15.5 (52)	1.2 (0)	16.0 (57)	0	14
	FE	14.0 (92)	1.1 (83)	15.9 (2)	5	0
	EX	14.5 (68)	1.4 (39)	12.0 (30)	71	16
Pollock	CE	8.8 (50)	0.52 (0)	16.5 (49)	0	9
	FE	8.6 (103)	0.50 (95)	16.5 (0)	2	0
	EX	8.7 (75)	0.63 (49)	12.4 (26)	79	12
Perch	CE	1.7 (22)	0.12 (0)	22.5 (22)	0	0
	FE	1.7 (82)	0.12 (88)	22.5 (0)	0	0
	EX	1.7 (51)	0.14 (50)	19.7 (14)	84	0

[a] Yield and stock biomass levels are expressed as a percentage of the virgin stock biomass x_0^*. Percentage CV in parentheses.

[b] Units for measuring effort are scaled within the model by the catchability coefficients and in this table are only useful for within-fishery policy comparisons. Percentage CV in parentheses.

[c] Percentage of years that minimum effort is employed. In the case of the FE policy, minimum effort is zero; that is, the entry is the percentage of years that the fishery is closed.

[d] Measured in percentage of years.

mental CV was set at 100% and the stock-recruitment shape parameter ρ was set at 0.25 in all three fisheries. The results in Table 4.9 indicate that long-term yields are very similar under all policies, with the greatest deviation being in the case of anchovy. In this fishery, the fixed-escapement policy results in 7% less yield when compared with the constant-effort policy. Note, however, that x_0^* is not necessarily the stock value that maximizes the long-term average yield over the class of possible fixed-escapement policies (see footnote on p. 172). The optimal fixed-escapement policy can be sought using Monte Carlo simulation techniques, and the maximum yield value obtained may even exceed MESY (see the case of widow rockfish discussed below).

The results presented here suggest that as long as a harvesting policy is designed around MESY conditions, long-term yield is relatively unaffected by the policy in question. The socio-economic implications of the various policies, however, are vastly different, as are the associated risks. For example, fixed-escapement closes the fishery down, or at least substantially reduces yield when stock biomass is low. In the case of the anchovy, the fixed-escapement policy closed the fishery on average once every 20 years. Such closures may protect the stock from a possible collapse, but the high variability in annual catch and hence income for participants in the fishery could be economically disastrous. The constant-effort policy, on the other hand, reduces the variance in annual catch by pushing more variance into the stock biomass levels. This may be satisfactory, provided the stock is afforded some measure of protection when biomass levels are low. The exploitative policy may be the easiest to justify politically, since fishermen are allowed to take advantage of the stock when it is strong and are otherwise allowed to fish at MESY effort levels. However, the average stock biomass levels are reduced by 10% to 25% when compared with the other two policies so that the risk of the fishery collapsing is substantially increased.

Getz, Francis, and Swartzman (1987) selected the *ad hoc* value $\frac{\hat{x}_0}{2}$ as an indicator of poor stock condition. From the last column in Table 4.9 we see that on average the pollock stock drops below $\frac{\hat{x}_0}{2}$ 9% of the time under an MESY constant-effort policy and approximately 11% of the time under an exploitative policy.

In summary, these results suggest that uncertainty in estimating MESY is more influenced by the shape of the stock-recruitment relationship, especially the rate at which recruitment falls off at low stock levels, than it is on the level of environmental variability associated with the recruitment process. Further, since fundamentally different policies are often indistinguishable on the basis of yield, the "best" policy must be determined by balancing yield and effort variability against stock variability and the probable risk of the fishery collapsing.

4.5.6   Optimal Harvesting of Widow Rockfish

Widow rockfish is a species off the west coast of the United States that has been heavily exploited only since 1980. In 1980–1983, landed catch averaged around 25,000 metric tons but declined to 9,000 metric tons by 1985. This decline is attributed not only to reduction in stock size, but also to restrictions on catch (quotas) by the U.S. Pacific Fisheries Management Council. In response to developments in this fishery, an analysis of optimal harvesting strategies was carried out by Hightower and Lenarz (in press).

They used data to bound the natural mortality rate as an age-independent constant lying somewhere between 0.15 and 0.20. Their analysis includes both extremes, but as results are qualitatively the same for both cases, we will refer only to the case $\alpha_i = 0.20$, $i = 1, \ldots, n$. They employed equations (4.22) as their model, although they related recruitment to stock levels four years rather than one year earlier. Thus, in their model, the first equation in (4.22) has the form

$$x_r(k+1) = l_r x_0(k - r + 1)\psi\big(x_0(k - r + 1)\big),$$

where $r = 4$ (cf. equations (3.8) and $i = r, \ldots, n - 2$). The parameters they used are listed in Table 4.10.

Hightower and Lenarz took the same approach as Ruppert *et al.* (1985); that is, they used relationship (4.30) to define their adaptive quota determining policy and searched for the parameters q and $\tilde{x}_0$ that would maximize the long-term yield for the two cases $\rho = 0.5$ and $\rho = 1.0$. The numerical algorithm they use (which they describe in some detail) is based on the algorithm formulated by

TABLE 4.10. Parameter values for the U.S. Pacific widow rockfish (*Sebastes entomelas*) fishery.

Number of age-classes	10 ($n = 13$)
Age at recruitment	$r = 4$
Length of harvesting season	$\bar{t} = 1$
Natural mortality parameters	$\alpha_i = 0.2$, $i = 4, \ldots, 13$
Weight-at-age[a]	400, 500, 600, 700, 900, 1000, 1100, 1200, 1300, 1600
Fecundity values[b]	0, 75, 102, 490, 711, 950, 1067, 1164, 1300, 1600
Stock-recruitment function[c]	$x_1 = \dfrac{27.4 x_0}{x_0 + 17{,}359}$
Relative catchability coefficients[d]	0.04, 0.18, 0.49, 0.72, 0.88, 1.00, 1.00, 0.84, 0.61, 0.61
Initial condition[e]	30.9, 25.1, 2.2, 7.9, 15.6, 8.3, 1.1, 0.8, 0.4, 8.7

[a] w_i listed by age $i = 4, \ldots, 13$; units are grams.

[b] b_i listed by age $i = 4, \ldots, 13$; units are grams, that is, x_0 is stock biomass measured in grams.

[c] The units are millions of individuals; the lognormal variance associated with this function is estimated to be $\sigma^2 = 1.04$.

[d] q_i listed by age $i = 4, \ldots, 13$.

[e] The units are millions of individuals; the values were obtained by estimating the structure of the stock at the beginning of 1986.

Ruppert *et al.* (1984). To obtain mean harvest estimates, they used a 65-year simulation period ($5 \times n$) replicated 100 times using Monte Carlo simulation but used only the values from the last thirty-three harvests to avoid the influence of the initial conditions. They also calculated the recommended 1986 yield obtained from each optimal policy. The results are listed in Table 4.11.

Since the value $\tilde{x}_0 = 20{,}707$ for the fixed-escapement policy that maximizes yield is half the MESY stock level $\hat{x}_0 \approx 40{,}000$, one would not expect the fishery to close too often. The negative value of $\tilde{x}_0$ for the fixed-escapement policy that maximizes the log-yield criterion implies that

TABLE 4.11. Yield results for the optimal policies for harvesting widow rockfish. After Hightower and Lenarz (in press).

Policy	Optimal parameters[a]	Mean yield[b]	1986 Quota[c]
Constant effort	$\hat{v} = 0.25$ $\hat{x}_0 \approx 40,000$	8,630(19)	8,040
	Yield criterion		
Fixed-escapement $(\rho = 1.0)$	$\tilde{q} = 0.67$ $\tilde{x}_0 = 20,707$	8,800(68)	14,010
Compromise fixed-escapement and constant catch $(\rho = 0.5)$	$\tilde{q} = 3.66$ $\tilde{x}_0 = 26,960$	8,820(69)	13,950
	Log-yield criterion		
Fixed-escapement $(\rho = 1.0)$	$\tilde{q} = 0.218$ $\tilde{x}_0 = -3,203$	8,520(49)	9,750
Compromise fixed-escapement and constant catch $(\rho = 0.5)$	$\tilde{q} = 1.73$ $\tilde{x}_0 = 11,075$	8,400(40)	9,530

[a] See equation (4.30); the units associated with x_0 are metric tons.

[b] Units are metric tons; the percentage coefficient of variation associated with the yield is given in parentheses.

[c] The catch determined by the policy assuming the initial conditions listed in Table 4.10; the units are metric tons.

the fishery will hardly ever close under this policy. The two yield maximizing policies had mean yields that were approximately 5% greater than the log-yield maximizing policies but coefficients of variation were roughly 50% greater; that is, the trade-off is to marginally improve the yield at a substantial increase in annual yield variance. As could be expected from the deterministic analysis in Section 4.4, the yield quotas that are set by different policies differ significantly more than the long-term averages. For example, we see from Table 4.11 that the yield-

187

maximizing policies set quotas for 1986 that are around 50% higher than the log-yield policies.

Although Hightower and Lenarz (in press) do not explicitly indicate how their policies performed in terms of catch per unit effort (CPUE), they do remark that their fixed-escapement policies maintain higher CPUEs than their constant-effort. A comparable result was also found by Swartzman *et al.* (1983) in their analysis of harvesting Pacific whiting using various constant-effort and fixed-escapement policies. This supports Hightower and Lenarz's conclusion that one cannot simultaneously obtain high yields, low yield variability, and high CPUE—there is always a trade-off.

Finally, Hightower and Lenarz were able to conclude for widow rockfish that, in the case of log-yield maximization, the optimal harvesting policy was very close to a constant-effort harvesting policy while, in the case of yield maximization, the optimal harvesting policy was intermediate between constant-effort and fixed-escapement.

4.5.7 Rational Harvesting of Anchovy

The case studies we have so far discussed in this section have focused on the comparative behavior of yield and stock variables under qualitatively different harvesting strategies. Although we have discussed questions relating to risk associated with collapse of a fishery, none of the studies treat this problem in any comprehensive fashion. This question deserves more attention, as provided by Bergh and Butterworth (1987) in their analysis of harvesting in the South African anchovy fishery.

Bergh and Butterworth use a model of the form (4.28), except they assume that harvesting corresponds to a pulse that occurs in the middle of the time interval $[k, k + 1]$; that is, each harvesting strategy determines a given number of individuals that are removed at time $k + \frac{1}{2}$. The stock recruitment curve they use is a ramp-type function in which recruitment depends linearly on stock up to a critical stock size $\check{x}_0$ and is constant beyond $\check{x}_0$. Hence their stock-recruitment relationship is a two-line approximation to a Beverton and Holt relationship (cf. Figure 3.1B). The value they chose for $\check{x}_0$ was 10% of the estimated unexploited stock biomass level. The exact values that they used for the other population parameters in their model are not given here. They differ to some extent from those listed in Table 4.1.

They compare three harvesting strategies (see Bergh and Butterworth, 1987, for more specific information):

- **Constant-catch.** The annual catch is fixed at a predetermined level $Y(k) = C$, but is reduced to $Y(k) = 0.95w_1(\frac{1}{2})x_1(k + \frac{1}{2})$ whenever $0.95w(\frac{1}{2})x_1(k + \frac{1}{2}) < C$ (recall that $w_1(\frac{1}{2})$ is the weight of an individual at age $\frac{1}{2}$).
- **Constant-proportion.** The annual catch $Y(k)$ is set as a proportion of $w_1(\frac{1}{2})x_1(k + \frac{1}{2})$.
- **Fixed-escapement analog**. The annual catch is set equal to zero unless the escapement target is estimated to be met, in which case the excess stock is harvested.

In addition, they looked at the problem of constraining the yield $Y(k)$ so that from one year to the next the change in harvest could not exceed a certain percentage ϵ; that is,

$$\frac{|Y(k + 1) - Y(k)|}{Y(k)} \leq e. \qquad (4.34)$$

The value of ϵ depends on economic considerations, specifically how easily capital and labor can be moved into or out of the fishery. Constraints, such as (4.34), are typically referred to as *flow constraints* in resource harvesting.

Bergh and Butterworth carried out two types of Monte Carlo simulation analyses, both of which included uncertainty associated with the recruitment process and with recruitment and stock-level measurement errors. In their

189

certainty-equivalence analysis they assumed that the best estimate of the state variables used to calculate $Y(k)$ were the levels predicted by a model with all parameters constant, except for recruitment level (median recruitment was multiplied by a lognormally distributed random variable). In their *myopic Bayes* analysis, they assumed that all the population parameters were drawn, as discussed in Section 2.4.3, from probability distributions. All simulations were carried out over a twenty-year period and the average annual yields and stock levels obtained over a large number of comparable runs were used to estimate the quantities evaluated in the analyses. The results obtained from both types of analyses are similar and only the results obtained from their certainty equivalence calculations are discussed here.

The comparative performances of risk versus yield for the above three harvesting strategies subject to a flow constraint parameter $\epsilon = 0.15$ are shown in Figure 4.9. The ordinate is the mean annual yield over a twenty-year harvesting period, while the abscissa is an estimate (using Monte Carlo simulations) of the probability of the stock falling below the level $\check{x}_0$ (10% of the mean unexploited stock level) at least once during the same twenty-year harvesting period. At the relatively low risk level of 10%, all three strategies result in comparable yields. At higher risk levels the escapement-analogous policy provides superior yields. In all three policies, the slope of the risk-yield curves are initially positive. They then reach a point corresponding to the maximum expected yield that it is possible to obtain, after which the slope of these curves becomes negative. Thus the trade-off between risk and yield reaches a point beyond which no additional yield can be obtained at an increased risk. The maximum possible yield is obtained using the escapement policy and the corresponding level of risk is 50%. This level of risk is almost certainly too high to tolerate provided the value selected

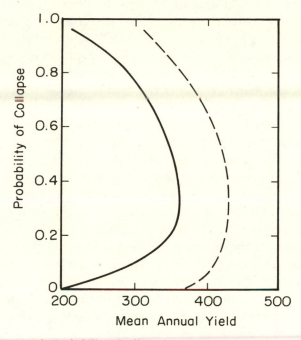

FIGURE 4.9. Probability of collapse in a 20-year harvesting period is plotted against mean annual yield (thousand metric tons) over the same period, where collapse is defined as the stock falling below 10% of its mean unexploited biomass level, for the harvesting policies: constant-catch (– – –); constant proportion (— · —); and fixed-escapement (——). (Data after Bergh and Butterworth, 1987.)

for $\check{x}_0$ (the level at which a collapse is defined to occur) is reasonable. Deciding what the appropriate level of $\check{x}_0$ is for a particular stock is virtually impossible, the only guide being some assessment of the levels at which stocks with comparable dynamics collapsed. The question of financial risk can also be brought into the analysis, as suggested by Bergh and Butterworth (1987).

A 15% flow constraint is restrictive and much higher expected yield levels can be obtained if these constraints are relaxed. For example, yields of 325(tmt), 405(tmt), and 460(tmt) are respectively obtained from the constant-proportion strategy at the 10% risk level for a 15%, 50%, and unrestricted flow constraint. Thus the management philosophy of trying to maintain small changes in catch from one year to the next can lead to a considerable reduction in long-term yield.

The above results assume that harvesting decisions are made on the basis of stock levels projected by the model. The coefficient of variation associated with these estimates can be reduced with the aid of annual surveys. With this reduced variance, one can expect to obtain a larger expected yield for a given risk level. The risk-yield curves associated with a 20% and a 40% coefficient of variation in recruitment for the constant-proportion strategy are illustrated in Figure 4.10. At the 10% risk level, the expected annual yields respectively are 435(tmt) and 325(tmt). This difference is quite substantial, but the increased value of the yield must be assessed against the cost of obtaining the required precision in the survey data.

4.6   AGGREGATED CATCH-EFFORT ANALYSIS

Many of the case studies discussed above assume that natural mortality is constant with age. It is important, however, to preserve age structure for those fisheries where gear selectivity questions are an issue. For fisheries, such as salmon, where gear selectivity is not an issue or where there is no clear evidence that catchability varies significantly with age, an aggregated age-structured model may be applicable. The one-dimensional form of the aggregated harvesting model (3.108) has substantial advantages over the n-dimensional model (4.22) when it comes to numerical estimation procedures and data analysis under stochastic conditions.

Our point of departure for the analysis in this section

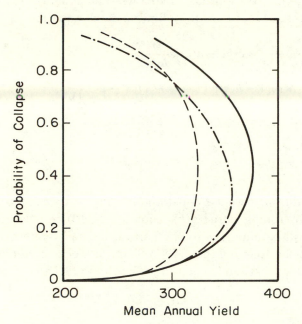

FIGURE 4.10. Probability of collapse in a 20-year harvesting period is plotted against mean annual yield (thousand metric tons) over the same period for the constant-proportion harvesting strategy (collapse is defined as the stock falling below 10% of its mean unexploited biomass level), for a coefficient of variation of $\sigma^2 = 0.2$ (broken line) and $\sigma^2 = 0.4$ (solid line) associated with the stock-recruitment process. (Data after Bergh and Butterworth, 1987.)

will be (3.108) under the assumption that the maturity and recruitment ages, i_0 and r respectively, are equal. The analysis would be more complicated if we did not make the latter assumption since then separate recruited stock and spawning stock variables would be required in the analysis.

In keeping with previous notation, the catch or yield taken over the time period $[k, k + 1]$ will be denoted by Y_k; that is, from (3.110) with $i_0 = r$ it follows that

$$Y_k = h(k)sx_0(k)/b.$$

193

Using this relationship, it follows that equation (3.108) reduces for $i_0 = r$ to the following catch equation:

$$\frac{Y_{k+1}}{h(k+1)} = c_1[1 - h(k)]s\frac{Y_k}{h(k)}$$
$$+ c_2[1 - h(k)][1 - h(k-1)]s^2\frac{Y_{k-1}}{h(k-1)}$$
$$+ bw_r[1 - h(k)]s\phi\left(\frac{Y_{k+1-r}}{h(k+1-r)}\right)$$
$$+ c_2bw_{r-1}[1 - h(k)][1 - h(k-1)]s^2\phi\left(\frac{Y_{k-r}}{h(k-r)}\right).$$
$$(4.35)$$

In addition, if we use equation (4.14) to relate the harvest $h(k)$ to fishing effort $v(k)$, equation (4.13) to relate the survival parameter s to the natural mortality rate α and the delay form of the Deriso-Schnute stock-recruitment function, that is (cf. equations (3.14) and (3.15)),

$$\phi(x_0) = l_r x_0(1 - \beta\gamma x_0)^{1/\gamma},$$

then equation (4.35) becomes the following catch-effort equation:

$$\frac{Y_{k+1}}{1 - e^{-qv(k+1)\bar{t}}} = c_1 e^{-\left(\alpha + qv(k)\bar{t}\right)}\frac{Y_k}{1 - e^{-qv(k)\bar{t}}}$$
$$+ c_2\left(e^{-\left[2\alpha + q\left(v(k) + v(k-1)\bar{t}\right)\right]}\right)\frac{Y_{k-1}}{1 - e^{-qv(k-1)\bar{t}}}$$
$$+ \left[bw_r e^{-\left(\alpha + qv(k)\bar{t}\right)}\frac{Y_{k+1-r}}{1 - e^{-qv(k+1-r)\bar{t}}}\right.$$
$$\left.\times\left(1 - \beta\gamma\frac{Y_{k+1-r}}{1 - e^{-qv(k+1-r)\bar{t}}}\right)^{1/\gamma}\right]$$
$$+ \left[c_2bw_{r-1}\left(e^{-\left[2\alpha + q\left(v(k) + v(k-1)\right)\bar{t}\right]}\right)\frac{Y_{k-r}}{1 - e^{-qv(k-r)\bar{t}}}\right.$$
$$\left.\times\left(1 - \beta\gamma\frac{Y_{k-r}}{1 - e^{-qv(k-r)\bar{t}}}\right)^{1/\gamma}\right].$$
$$(4.36)$$

For a given season length $\bar{t}$, equation (4.36) can be used to predict the next catch from previous catch and effort data, specifically Y_{t-k}, Y_{t-k+1}, Y_{t-1}, Y_t, $v(t-k)$, $v(t-k+1)$, $v(t-1)$, and $v(t)$, in terms of the current effort $v(k+1)$. This assumes that the eight model parameters α, β, γ, c_1, c_2, q, w_r, and w_{r-1} are known. If some or all of these parameters are not known, then, as discussed in some detail in Schnute (1985), a time series of catch-effort data may be used to estimate the most likely values for the unknown parameters (for further discussion, see Schnute, 1985). Schnute (1987) also presents equations comparable to (4.36) for size-structured fish populations.

As in Section 4.3.1, yield-effort curves can be constructed for the aggregated model under equilibrium conditions. From equation (4.36) it follows that if an equilibrium exists corresponding to the constant-effort level v, it will satisfy the equation

$$Y(v) = \frac{1 - e^{-qv\bar{t}}}{\beta\gamma} \left[1 - \left(\frac{e^{\alpha+qv\bar{t}} - c_1 - c_2 e^{-(\alpha+qv\bar{t})}}{bw_r + bw_{r-1}e^{-(\alpha+qv\bar{t})}} \right)^{\gamma} \right].$$

Clearly $Y(0) = 0$, but the general form of $Y(v)$ depends very much on the values of the different parameters. There are too many parameters to undertake a comprehensive discussion of all the cases that can arise. In most cases, yield-effort curves have the same basic form as those presented for age-structured models.

Equation (4.35) can be made stochastic in a number of different ways. For example, one could assume that stochasticity enters the fishery through fluctuations in the natural mortality coefficient α, and/or errors in the measurement of the effort levels $v(k)$ (or equivalently uncertainty in the catchability coefficient q, and/or errors in the assessment of the weight parameters w_{r-1} and w_r, and/or assesment of the recruitment function

ϕ). The particular aspect that should be emphasized depends on the fishery under consideration. With the addition of stochastic elements, both analytical procedures for finding the approximate moments of the catch variable and Monte Carlo procedures for numerically generating these moments and other statistics can be undertaken as outlined for the higher-dimensional stage-structured models.

4.7 MULTISPECIES MULTIPARTICIPANT FISHERIES

4.7.1 Introduction

Every species of fish in a marine environment is part of a complex ecosystem. Two species that compete for food resources or two species involved in a prey-predator interaction may each support a different fishery. Thus fisheries may indirectly interfere with one another because of the ecological interactions between the species they each harvest. A number of multispecies fisheries studies have looked at the problem of including prey-predator interactions in their analyses (for example, see the selection of papers in Mercer, 1982; or see Butterworth *et al.*, 1988, for a discussion of simultaneously harvesting seals and their fish prey). It is rather difficult, however, to estimate predation and competition rates, while it is virtually impossible to characterize such critical density-dependent and seasonal processes as prey switching and preference. Furthermore, the number of species in the food chain associated with a commercially exploited fish stock can be overwhelming.

Pacific whiting, for example, not only supports its own fishery, but is a key species in the ecology of several other North American West Coast fisheries. Pacific whiting is an important item in the diet of a number of rockfish, flatfish, and roundfish species, and provides a prey item for several species of marine mammals (Rexstad and Pikitch, 1986). Pacific whiting itself feeds on the smaller individuals of those species that it is vulnerable to when young. It also feeds on numerous species of crustaceans.

As we have seen in a number of the single-species fisheries examined in previous sections, age-specific mortality rates are usually estimated as having the same age-independent constant value (see Tables 4.1, 4.6, 4.8, and 4.10. Thus if we are unable to differentiate between mortality rates with respect to age, we cannot partition mortality rates among the various predators with any reasonable level of accuracy.

At the fisheries level, there is a more direct form of multispecies interaction when several stocks of fish are exploited concurrently in time and space. Pope (1979) refers to these as *technological interrelationships*, although a finer level of distinction can be made. For example, species may be mixed at the gear level (species school together or, at least, are mixed at a fine spatial level). On the other hand, the landed catch of a single boat may be mixed because different species are caught as they are encountered at different times during a particular trip. The distinction is important because of the so-called *by-catch* problem, which is catching unwanted species in the gear with the targeted species. Thus a mix of different species landed by a boat may be due to the by-catch problem, or may simply be from catching different species at different times during the same trip. Only in the latter case can the proportion of different species in the total catch be controlled.

Some by-catch species may be valueless and a nuisance to sort out. Others may be protected (for example, the quota associated with a particular species may have already been exceeded for the season) so that the fisherman will have to discard them. Such discarding can be problematic for protected, overexploited stocks because the discarded fish are usually dead and cannot contribute to the future production of that stock. Also, discarding is not reflected in the catch data, thereby resulting in biased estimates of population parameters. Although very little is know about discarding rates, current multispecies studies are beginning to address the discard problem.

Another aspect of multispecies harvesting is the number of distinct fisheries in which a particular species may feature (Murawski *et al.*, 1983; Murawski, 1984). For example, a species may be vulnerable to different types of gear used by different boats operating in the same area even though all individuals belong to the same stock (i.e., all individuals migrate to the same area to spawn as a unit). Also, a species may be distributed during part of the season over a number of geographic areas where it is vulnerable to several spatially separated fleets of boats. The latter may include fleets from different nations subject to different sets of regulations. In this context, analysis of harvesting strategies, as is discussed below, moves into a dynamic game formulation.

In keeping with our top-down modeling approach, we will not attempt to address ecological species interactions. Here we will focus on extending the single-species models of the previous sections to link several such models through technological interrelationships. This approach has been used to address questions relating to the setting of quotas in fisheries that exploit multispecies assemblages. The advantage of a technology-linkage multispecies analysis over several independent single species analyses is that species are often linked at the gear level (by-catch); it is uneconomical for fishermen to pass up the opportunity of catching more than one species per trip; and fluctuations in species abundance may help to smooth out the availability of fish as a whole.

4.7.2   Technological Links

We generalize our previous approach by defining x_{ij} as the number of individuals in age class i of species j, where $i = 1, \ldots, n_j$ for each j (i.e., n_j is the number of age classes for species j) and $j = 1, \ldots, m$ (i.e., in this section m denotes the number of species in the assemblage). The vector $\mathbf{x}_j = (x_{1j}, \ldots, x_{n_j j})'$ is used to represent the state of species j, which still satisfies the basic system of transition equations (4.22), except the index j is now used to indicate that this system of equations applies to the jth species. Thus for $j = 1, \ldots, m$, the jth species satisfies the system of equations,

$$x_{1j}(k+1) = s_{0j}x_{0j}(k)\psi_j\big(x_{0j}(k)\big),$$

$$x_{i+1\,j}(k+1) = e^{-\left(\alpha_{ij}+q_{ij}v(k)t\right)}x_{ij}(k), \quad i = 1, \ldots, n_j - 2$$

$$x_{n_j j}(k+1) = e^{-\left(\alpha_{n_j-1\,j}+q_{n_j-1\,j}v(k)t\right)}x_{n_j-1\,j}(k)$$

$$+ e^{-\left(\alpha_{n_j j}+q_{n_j j}v(k)t\right)}x_{n_j j}(k), \tag{4.38}$$

where $x_{0j}(k) = \sum_{i=1}^{n_j} b_{ij}x_{ij}(k)$ is the spawning stock variable or egg index, and the effort level $v(k)$ is interpreted as the effort level applied to the total species assemblage during the kth time interval. The actual partition of effort among species and age classes is determined by the extended set of catchability coefficients q_{ij}.

These catchability parameters lie at the heart of multi-species technological interaction problems. One can use historical catch data to calculate the various proportions of individuals by species and age and use these values to estimate the relative values of the parameters q_{ij}. For example, recalling expression (4.15), the biomass of individuals harvested from age class i of species j (where w_{ij} is the average weight of each individual) in the kth season is

$$C_{ij}(k) = \int_k^{k+1} v(k)w_{ij}q_{ij}(k)x_{ij}(t)dt. \tag{4.39}$$

If x_{ij} remains relatively constant over the period $[k, k+1]$, then it follows that

$$q_{ij}(k) \approx \frac{C_{ij}}{v(k)x_{ij}(k)}.$$

The difficulty with this approach is that $x_{ij}(k)$ itself must be estimated independently of $C_{ij}(k)$ and $v(k)$. Further, this approach ignores the fact that $v(k)$ is hardly ever applied homogeneously to the multispecies assemblage.

In general, effort has a structure that reflects the targeting behavior of various boats. If a boat is targeting on species j, then the probability that nontarget species appear in the catch will not be reflected by the mix of species in the catch history. Only by-catch information (which is not readily available since it is obscured in landing data when boats target on several species during one trip) can be used to estimate selectivity coefficients associated with targeting. Define π_r to be the proportion of harvesting effort spent targeting on species r, $r = 1, \ldots, m$, where $\sum_{j=1}^{m} \pi_j = 1$. Then we need to estimate selectivity coefficients q_{ijr} which represent the relative proportion of individuals of age i in the j-species that are vulnerable to harvesting by gear that is targeting on species r. In this case we replace the terms in $q_{ij}v(k)$ in system (4.38) with the right-hand side of

$$q_{ij}v(k) = \sum_{r=1}^{m} \pi_r q_{ijr} v_r(k). \qquad (4.40)$$

If individuals in species j and r are distinct at a coarse enough level of spatial resolution then we can expect that $q_{ijr} = 0$. Thus a knowledge of how species distribute themselves in an area can be used to reduce the actual number of selectivity coefficients that need to be estimated. Note also that once the total effort can be split into a number of different targeting effort levels v_r, then each v_r may be applied over a different portion of the harvesting season. Thus we will assume that v_r is applied over $[t_{1r}, t_{2r}]$ where $0 \leq t_{1r} < t_{2r} \leq 1$.

One could also take the resolution one step further to an individual boat level. The relative activity parameters π_r in expression (4.10) can be obtained by summing over the amount of time spent by each individual boat on a particular activity. This approach is useful if we need to distinguish between the operating costs and efficiencies of different classes of boats participating in different activities. The details of this approach are given in Getz *et al.* (1985).

4.7.3 Sustainable Yields, Quotas, and Discarding

Although, as we have seen, sustainable yield has no well-defined meaning in stochastic fisheries, it is still instructive to consider sustainable yields in the context of managing individual species and in managing these species as an assemblage. Assume that a set of weight-at-age constants w_{ij} is available for species $j = 1, \ldots, m$. Define

$$\mathcal{Q} = \{q_{ijr} | i = 1, \ldots, n_j, \; j, r = 1, \ldots, m\}$$

and assume that the times t_{1r}, $r = 1, \ldots, m$, are selected *a priori* (according to some biologically or economically related criterion). Then it can be shown that the equilibrium yield from the jth species corresponding to the choice of effort levels $\hat{\mathbf{v}} = (\hat{v}_1, \ldots, \hat{v}_m)'$ and lengths of harvesting periods characterized by $\mathbf{t}_2 = (t_{21}, \ldots, t'_{2m})$ is given by (cf. expressions (4.15), (4.18), (4.39), and (4.40))

$$Y_j(\hat{\mathbf{v}}, \mathbf{t}_2, \mathcal{Q}) = \sum_{r=1}^{m} \left[\hat{\pi}_r v_r \right.$$

$$\times \sum_{i=1}^{n} \frac{w_{ij} q_{ijr} \hat{x}_{ij} \left(e^{-\alpha_{ij} t_{1r}} - e^{-\left[\alpha_{ij} t_{2r} + \sum_{r=1}^{m} \pi_r q_{ijr} \hat{v}_r (t_{2r} - t_{1r}) \right]} \right)}{\left(\alpha_i + \sum_{r=1}^{m} \pi_r q_{ijr} \hat{v}_r \right)} \left. \right].$$

$$\tag{4.41}$$

Let Y_j^* represent the MSY value of species j harvested in isolation from the rest of the multispecies assemblage. This can be achieved only if $q_{ijr} = 0$ whenever $r \neq j$, that is, if we are able to target without any by-catch occurring. Then the maximum sustainable biomass from the assemblage is

$$Y^* = \sum_{j=1}^{m} Y_j^*. \tag{4.42}$$

(Note that we can interpret Y as value rather than yield, if the constants w_{ij} are themselves value-specific rather than weight-specific numbers.)

If the jth species cannot be harvested in isolation from the multispecies assemblage, then, in general, Y^* is no longer achievable. The maximum yield

$$\tilde{Y}(\mathcal{Q}) = \sum_{j=1}^{m} Y_j(\mathbf{v}^*, \mathbf{t}_2^*, \mathcal{Q}), \tag{4.43}$$

obtained at the optimum effort level $\mathbf{v}^*$, must necessarily satisfy $\tilde{Y}(\mathcal{Q}) \leq Y^*$ because of the loss of ability to independently harvest each species. This situation is analogous to the relationship between the USY and MSY in the single species case, as discussed in Section 4.3.

Under equilibrium conditions one may choose to regulate fishing effort to correspond to $\mathbf{v}^*$. Recruitment to the various stocks will fluctuate in practice. Regulations, or the setting of quotas, need to be responsive to these fluctuations so that advantage can be taken of strong year classes in particular stocks, and those stocks that are much weaker than normal can be protected. Setting quotas for individual stocks is useless since, as soon as the quota is reached, any individuals belonging to that stock that are subsequently taken as by-catch will be discarded at sea (i.e., individual quotas can only be set for stocks that do not appear as by-catch in the operation of the fishery or fisheries).

Suppose at time $t = 0$, the jth stock is in state $\mathbf{x}_j(0)$, $j = 1, \ldots, m$; then one can consider the T-year planning horizon problem of maximizing

$$J_T = \sum_{k=0}^{T-1} \sum_{j=1}^{m} Y_j(\mathbf{v}(k), \mathbf{t}_2(k), \mathcal{Q}),$$

subject to systems of equations (4.38) holding for each species, and specified final time conditions corresponding to the optimal equilibrium effort level $\mathbf{v}^*$ (cf. Problem 5, p. 155).

As discussed in Section 4.5, fish stocks are stochastic and it is difficult to accurately estimate the state of most stocks at any point in time. The results of the single-species analysis suggest that the dynamically suboptimal best effort policy $\mathbf{v}^*$ provides a reasonable basis for regulating effort as long as variation in stock levels can be tolerated in terms of the risk associated with the collapse of a particularly weakened stock. Because of the by-catch and discard problem, it is no longer possible to regulate each individual stock's estimated escapement level independently. The only parameters that can be selected independently are the effort levels $\mathbf{v}(k)$ and the lengths of the harvesting seasons $\hat{\mathbf{t}}_2$. One can follow the methods discussed in Section 4.5.4, to define v_r, $r = 1 \ldots, m$, adaptively, and estimate a set of parameters that maximizes the long-term average of the yield $Y_k = \sum_{j=1}^{m} Y_j(\mathbf{v}(k), \mathbf{t}_2(k), \mathcal{Q})$. One scheme is to determine the optimal parameters for v_r (cf. equation (4.30)) defined in terms of an aggregated assemblage stock variable,

$$\mathbf{x}_0(k) = \sum_{j=1}^{m} \omega_j x_{0j}(k),$$

(ω_j are appropriately selected weighting constants) and species-specific parameters c_{1r}, c_{2r}, c_{3r}, and $\tilde{v}_r$ (the first two are stock-comparison parameters) as follows:

$$v_r(k) = 0, \text{ when } \mathbf{x}_0(k) \le c_{1r};$$

$$\left.\begin{array}{l} v_r(k) = \tilde{v}_r \\ t_{2r} = \frac{\mathbf{x}_0(k) - c_{1r}}{c_{2r} - c_{1r}} \bar{t}_{2r} \end{array}\right\} \text{ when } c_{1r} < \mathbf{x}_0(k) \le c_{2r};$$

$$\left.\begin{array}{l} v_r(k) = \tilde{v}_r \left(\frac{\mathbf{x}_0(k) - c_{1r}}{c_{2r} - c_{1r}} \right)^{c_{3r}} \\ t_{2r} = \bar{t}_{2r} \end{array}\right\} \text{ when } \mathbf{x}_0(k) > c_{2r}, \qquad (4.44)$$

where the vector $\bar{\mathbf{t}}_2$ represents the maximum time that the season associated with the rth targeting activity can remain open.

If many species are involved, the computational time required to estimate the parameters associated with all r targeting activities defined in (4.44) may be excessively large and some simplifications may need to be made.

4.7.4 U.S. West Coast Groundfish Resource

A number of species of rockfish (*Sebastes* complex), flat-fish, and roundfish are harvested off the west coast of the United States (California, Oregon, Washington). Catches are taken by trawlers primarily operating out of about eight coastal ports. This groundfish fishery or complex of fisheries is regulated by the U.S. Pacific Fisheries Management Council on advice from various statutory committees and organizations. Catch limitation regulations are difficult to enforce because of the problem of discarding at sea those fish that are unlawful to land. Thus in analyzing the management of this resource, it is important to include discarding.

The distribution of fish (species and numbers) and boats (number and type) is spatially heterogeneous, and subcomponents of this resource can be focused upon for the purposes of analysis, provided one does not lose sight of the links between these subcomponents. Pikitch (1987) has

204

TABLE 4.12. Parameter values for the southern Oregon sole fishery.

	Dover sole		English sole		Petrale sole	
	M^a	F^a	M	F	M	F
Dynamic parameters[b]						
r	5	5	1	1	1	1
α	0.15	0.15	0.26	0.26	0.25	0.25
q	15.7	15.7	22.0	22.0	45.0	45.0
x_1	10.9	13.6	0.8	5.4	3.8	2.3
Growth parameters[c]						
$\bar{w}$	595	1130	391	764	1571	1486
c_1	0.86	0.81	0.76	0.90	1.02	0.68
c_2	0.094	0.094	0.256	0.265	0.202	0.21
c_3	2.89	2.97	3.01	3.40	3.38	3.28
c_4	13.4	10.1	7.8	2.2	4.0	3.0
Economic parameters[d]						
p	0.55	0.55	0.75	0.75	1.65	1.65
l_{min}	34	34	31	31	33	33

[a] M = male, F = female.

[b] The number of recruits x_1 (in millions) is assumed constant, i.e., independent of stock. The number of age classes used in the analysis was taken as 40 for all species, and r is the age at recruitment. The mortality parameters α and catchability coefficients q are assumed to be constant with respect to age and differ only with respect to sex and species. The length of the harvesting season was taken as $t = 1$.

[c] These are the von Bertalanffy parameters for the weight function $w(t)$ described in equation (4.3). The units are grams. The parameter c_4 is used to relate length-at-age, $l(t)$, with weight using the formula $l(t) = \left(\dfrac{w(t)}{c_4 \times 10^{-3}} \right)^{1/c_3}$. The unit for length is centimeters.

[d] The parameter p is the 1985 ex-vessel price quoted in \$/kg. The parameter l_{min} is the minimum marketable size in centimeters.

analyzed an otter trawl segment of this fishery, operating out of Coos Bay in southern Oregon, and employing "sole gear" to catch primarily English sole (*Parophrys vetulus*), Dover sole (*Microstomus pacificus*), and Petrale sole (*Eopsetta jordani*). She assumed that only technological interactions play a role since no strong ecological links (competition or predation) are apparent. She employed a model of the form outlined in Section 4.7.2, but assumed that recruitment in equation (4.38) is a constant for each species; that is, she essentially conducted a tri-species yield per recruit analysis.

The species specific biological and economic parameters used by Pikitch are listed in Table 4.12.

For each species, the growth characteristics of the different sexes are sufficiently different to warrant their separation. Age-specific catchability coefficients were calculated using the species-specific catchability coefficients listed in Table 4.12 multiplied by an age selectivity factor determined from gear selectivity ogives. These ogives determine selectivity with respect to length which is related to age by the functions that depend on the growth parameters listed in Table 4.12. The ogives for the smallest and largest of the four mesh sizes considered in the analysis (102 mm, 114 mm, 127 mm, and 140 mm) are illustrated in Figure 4.11. The species-specific catchability coefficients were calibrated to correspond to an equilibrium annual effort level of 9,566 hours (observed average from 1971 to 1974). Assuming equilibrium conditions, Pikitch estimated the number of recruits, x_1, in Table 4.12 by running the model separately for each species at the equilibrium effort level, and calibrating to obtain observed average catch levels during 1971–1974. Included in Pikitch's analysis were the calculation of the species-specific yields $Y_j(\hat{\mathbf{v}}, \mathbf{t}_2, \mathcal{Q})$ and also the *total net revenue J*, where the latter is calculated from the species specific ex-vessel prices p_j (see Table 4.12 for the values used) per marketable unit biomass of species j and a cost per unit fishing effort c using the formula

$$J = \sum_{j=1}^{m} p_j Y_j - cv,$$

where the effort variable v is measured in trawling hours per season. Pikitch used a value of $c = 36.42$ U.S. dollars per trawling hour (for the derivation of this value see Pikitch, 1987). Only fish in the catch above minimum market length (l_{min}, Table 4.12) were used to calculate total net revenue, while it was assumed that a proportion of the fish as determined from the discard curves (reverse ogives), illustrated in Figure 4.12, were discarded at sea, and thus could not be included in yield or revenue calculations.

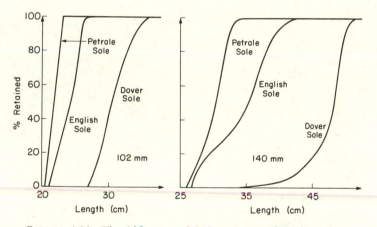

FIGURE 4.11. The 102 mm and 140 mm gear selectivity ogives for the southern Oregon sole fishery. (Adapted from Best, 1961.)

For this fishery, Pikitch calculated both the maximum sustainable yield $\tilde{Y}$ of the technologically linked stocks (see expressions (4.41) and (4.43)), and compared it with theoretical maximum Y^* (see expression (4.42)) which could be obtained only if the stocks were not technologically linked. Values of approximately 2700 and 2100 metric tons per annum were respectively obtained for Y^* and $\tilde{Y}$ so that 20%–25% of the yield is lost because of the technological links. The MSY yield, $\tilde{Y}$, was obtained using the 127 mm mesh. The proportion of Petrale sole in the catch was extremely low, because Petrale sole is more easily caught than the other two species (see Figure 4.11) so that the

207

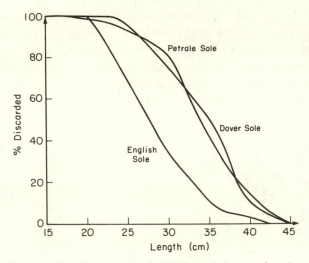

FIGURE 4.12. Percentage of fish discarded, as a function of length, in the southern Oregon sole fishery. (Adapted from Ten-Eyck and Demory, 1975.)

TABLE 4.13. Yield (metric tons) for the southern Oregon sole fishery.

Management policies	Mesh size (mm)			
	102	114	127	140
Maximum sustainable yield	1745	1942	2093	1529
Bionomic equilibrium yield[a]	1236	1648	2092	1275

[a] Net equilibrium revenue is zero.

Petrale stock was relatively overexploited at the assemblage MSY point. Since Petrale commands two to three times the price of the other two stocks (see Table 4.12), the maximum revenue solution includes a much greater proportion of Petrale in the catch (see Pikitch, 1987, for details). The best sustainable yields that could be obtained using 102 mm, 114 mm, and 140 mm mesh sizes were respectively 83%, 93%, and 73% of $\tilde{Y}$ (Table 4.13).

The MSR (net revenue) solution, using the price and

cost parameters discussed above, was obtained using a mesh size of 114 mm and corresponded to an annual return of $1.2 million. Net revenues within 97% of this could be obtained using 102 mm and 127 mm gear, but only 78% of this value could be obtained using 140 mm gear.

It is very difficult, however, to control effort in a fishery at MSY or MSR levels. Further, it is not necessarily desirable to apply MSY or MSR effort levels since so many other socio-economic factors, central to the fishery, are not included in the analysis. Some theories on the exploitation of open-access resources maintain that the resource will be exploited to the point where all net revenues are dissipated (i.e., costs and gross revenues are equal; see Gordon 1954; Clark 1976, 1985). This dissipation of revenues is referred to as the "tragedy of the commons" (Hardin, 1968), while the corresponding population level is termed the "bionomic equilibrium" (Clark, 1976).

If we compare maximum sustainable yields and yields obtained under the rent dissipation assumption for the different mesh sizes, as listed in Table 4.13, it is apparent that the MSY solution (127 mm maximum sustainable yield) is almost at the rent dissipation point. At the two mesh sizes smaller than 127 mm, maximum sustainable yield is obtained with relatively less effort than the 127 mm case so that additional effort brought in to dissipate the rent leads, especially in the case of 102 mm mesh size, to considerable biological overexploitation (in the 102 mm case, a 30% reduction in marketable yield is observed, while the discard rate, although not indicated, is bound to be much higher than that in the 102 mm maximum sustainable case). For the mesh size larger than 127 mm, the effort at the maximum sustainable solution is sufficiently high so that the net revenue is negative, and the reduction in effort to obtain the zero-revenue solution leaves that stock assemblage relatively unexploited (only 83% of the maximum sustainable yield is obtained). Thus the 127 mm gear appears to be ideal for managing a fishery in the absence of any effort regulations, since the bionomic equilibrium and MSY solutions practically coincide. Of course a sole owner could more profitably exploit the fishery using a smaller mesh size (MSR is obtained using 114 mm gear).

Pikitch also conducted a number of "sensitivity" studies to see how particular solutions are affected by relative changes in economic and recruitment parameters. She concluded that the most appropriate mesh size was more likely to depend on the formulation of the management question (namely, MSY, MSR, or bionomic equilibrium) than on economic or recruitment parameters. Obviously, a number of other important questions come up and Pikitch's study represents only a beginning to the problem of analyzing the U.S. West Coast groundfish fishery. The effects of a stock-recruitment relationship need to be assessed and the performance of gear types under dynamic (stochastic) conditions need to be evaluated. Other major species can be included (flatfish and roundfish), as can boats from several geographic areas. It is hard to know exactly where to draw the line and what the major factors are in the problem. As the geographic and socio-economic boundaries of the problem are broadened, other details— such as the effects of age structure—are blurred, and may even become unimportant.

4.7.5 U.S.-Canadian Pacific Whiting Fishery

Pacific whiting, *Merluccius productus*, also referred to as Pacific hake, consists of four major spawning stocks (Bailey *et al.*, 1982). The most abundant of these is the so-called "coastal stock" fished off the coasts of California, Oregon, Washington, and British Columbia. This stock spawns off the coasts of southern California and Baja California during January, and then migrates northwards to feed. The youngest individuals (juveniles) school in the most southerly location (just north of San Francisco Bay), with larger individuals schooling progressively north, the largest of which reach Vancouver Island in late May (Figure 4.13) Most of the stock remains in U.S. territorial waters where the Pacific whiting fishery is regulated by U.S. authorities. Once the larger individuals enter Canadian waters, however, they are exploited in fisheries regulated by Canada. Most of the fish are taken in joint-venture fisheries in which U.S. and Canadian boats in their respective territorial waters catch the fish and unload them onto foreign vessels for processing and marketing.

It is clear from the migratory pattern depicted in Figure 4.13 that the United States can destroy the Canadian fishery by harvesting large individuals before they enter Canadian waters. Such a strategy, however, would be to the detriment of the U.S. fishery, especially since larger fish are proportionally more fecund per unit weight than smaller fish. Thus both fisheries have a vested interest in maintaining a significant number of large fish in the spawning stock. If we regard the U.S. and Canadian regulators as players in a two-person game, where each player is trying to maximize the value of his payoff (for example, revenue or harvested biomass), then the best strategy for each player depends on whether the two parties cooperate and trust each other. A full discussion of game theory is beyond the scope of this book.[2] Here we only discuss the relevant issues as they arise for the problem on hand.

Hobart (1985) reviews the biology of the stock and the operation of the fishery. The material presented in this subsection is a summary of Swartzman *et al.*'s (1987) analysis of Pacific whiting as a binational fishery management problem. Recruitment for this stock is quite variable and appears to be correlated to sea-surface tempera-

---

[2] The reader is referred to Kaitala (1986) and Hämäläinen *et al.* (1984, 1985) for discussions on the application of differential game theory to fisheries management problems.

211

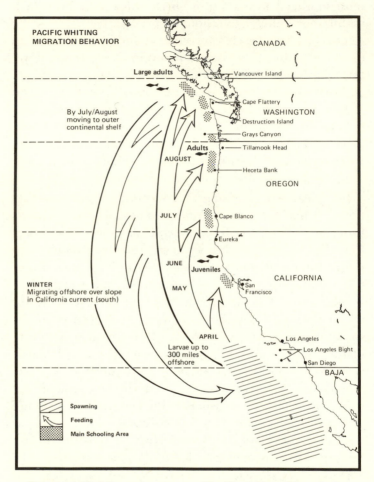

FIGURE 4.13. Migratory patterns of Pacific whiting. (From Bailey *et al.*, 1982.)

ture (as consequence of the effects of surface currents—Ekman transport conditions—and larval survival rates). To avoid Monte Carlo simulations but still obtain variance/covariance estimates, Swartzman *et al.* used the large noise stochastic recruitment model developed in Section

3.5.3 to model the natural stock dynamics (this approach allows a much more comprehensive management analysis to be undertaken because only one rather than 25 to 100 runs needs to be undertaken for each iteration of the harvesting algorithm). For the purposes of analysis, all age classes from recruitment at age 3 onwards were split each season into a proportion π_i that remained in the U.S. fishery and $(1 - \pi_i)$ that entered the Canadian fishery, $i = 3, \ldots, n$. The actual proportions used are listed in Table 4.14. They were assumed to be constant over time. Thus if $v^a(k)$ and $v^c(k)$ respectively denote the level of effort applied in the U.S. and Canadian fisheries and q_i^a and q_i^c are the corresponding catchability coefficients, then for $i = 1, \ldots, n-2$ (or $n-1$ if the truncation term is ignored), the transition equations have the form (cf. the second line of equations (3.90) and (4.22))

$$x_{i+1}(k+1) = \left(\pi_i e^{-\left(\alpha_i + q_i^a v^a(k)t\right)} + (1 - \pi_i)e^{-\left(\alpha_i + q_i^c v^c(k)t\right)} \right)$$
$$\times x_i(k). \qquad (4.45)$$

The corresponding covariance equation (cf. the last line in equations (3.90)) is:

$$\gamma_{i+1\,j+1}(k+1) = \left(\pi_i \pi_j e^{-\left(\alpha_i + \alpha_j + (q_i^a + q_j^a)v^a(k)t\right)} \right.$$
$$\left. + (1 - \pi_i)(1 - \pi_j)e^{-\left(\alpha_i + \alpha_j + (q_i^c + q_j^c)v^c(k)t\right)} \right)\gamma_{ij}(k). \qquad (4.46)$$

It is not clear how realistic it is to assume that the proportions π_i are independent of population density. A more realistic model, however, would require significantly more data and knowledge of migration behavior than is currently available.

The population parameters and catchability coefficients used in this study are also listed in Table 4.14. Note that the stock-recruitment function has the compensatory form introduced in equation (3.9). Also note that fish are recruited to the U.S. fishery at age three, but are recruited to the Canadian fishery only at age five ($q_3^c = q_4^c = 0.0$). Although a greater proportion of individuals in all age classes, except for the last age class, are available to the U.S. fishery, the older fish (especially the last three age-classes) are more vulnerable to fishing pressure (relatively larger catchability coefficients) in the Canadian fishery.

Consider the situation where the U.S. and Canadian players respectively select constant effort levels v^a and v^c. Then from equation (4.45), it follows by analogy in deriving expression (4.19) that the U.S. yield in season k is $Y_k^a = Y\big(v^a(k), \mathbf{q}^a, \boldsymbol{\pi}\big)$, where

$$Y\big(v^a(k), \mathbf{q}^a, \boldsymbol{\pi}\big) = v^a(k)$$
$$\times \sum_{i=1}^n w_i q_i^a \pi_i x_i(k) \left[1 - e^{-\left(\alpha_i + q_i^a v^a(k)\hat{t}\right)}\right] \Big/ \big(\alpha_i + q_i^a v^a(k)\big).$$

Similarly, the Canadian yield $Y_k^c = Y\big(v^c(k), \mathbf{q}^c, \mathbf{1} - \boldsymbol{\pi}\big)$, where the vector $(\mathbf{1} - \boldsymbol{\pi})$ has elements $(1 - \pi_i)$.

Swartzman *et al.* (1987) examined a number of scenarios. In one of these they assumed that the U.S. and Canadian players were cooperating to maximize the value J (defined below) of the game over a T-year planning horizon by respectively selecting an optimal sequence of harvesting effort levels $v^a = \big(v^a(0), \ldots, v^a(n-1)\big)'$ and $v^c = \big(v^c(0), \ldots, v^c(n-1)\big)'$, where

$$J(v^a, v^c) = \sum_{k=0}^{T-1} c_1\big(Y_k^a - c_2 v^a(k)\big) + (1 - c_1)\big(Y_k^c - c_2 v^c(k)\big).$$
$$(4.47)$$

Note that the constant c_1 reflects the relative value placed on net returns from the U.S. fisheries with respect to the Canadian fishery, and the constant c_2 reflects the cost/price ratio of effort and yield which is assumed to

TABLE 4.14. Parameter values for the Pacific whiting U.S./Canadian Fishery.

Number of age-classes	$n = 10$
Age at recruitment	$r = 3$
Length of harvesting season	$\bar{t} = 0.42$
Natural mortality parameters	0.27, 0.21, 0.20, 0.26, 0.36, 0.46, 0.56, 0.66, 0.76, 0.86
Weight-at-age[a]	433, 545, 663, 769, 855, 947, 1023, 1078, 1134, 1205
Fecundity values[b]	33, 66, 113, 136, 155, 177, 194, 208, 221, 239
Stock-recruitment function[c]	$x_1 = 470 \times 10^6 \left(1 - e^{-1.22 \times 10^{-14} x_0}\right)$
Relative U.S. catchability coefficients[d]	0.006, 0.020, 0.030, 0.042, 0.054, 0.099, 0.129, 0.150, 0.277, 0.277
Relative Canadian catchability coefficients[e]	0.000, 0.000, 0.052, 0.054, 0.078, 0.093, 0.177, 0.379, 0.762, 0.762
Proportion of individuals in the U.S. fishery[f]	1.00, 0.99, 0.92, 0.85, 0.78, 0.70, 0.63, 0.58, 0.52, 0.45

[a] w_i listed by age $i = 3, \ldots, 12$; units are grams.

[b] b_i listed by age $i = 3, \ldots, 12$; units are numbers of eggs per individual age i; i.e., x_0 is an egg index.

[c] This is the recruitment function averaged over years classified as either warm or cold. The units are individuals. The coefficient of variation associated with recruitment for warm years was 107% and for cold years 62% (see Swartzman et al. 1987 for further details).

[d] q_i^a listed by age $i = 3, \ldots, 12$.

[e] q_i^c listed by age $i = 3, \ldots, 12$.

[f] These values of π_i^c were calculated from w_i using the formula associated with Figure 10 in Francis et al. (1984).

be the same in the two fisheries (a more general criterion can just as easily be considered). In a cooperative situation, once the two players have negotiated an acceptable value for c_1, the optimal harvesting sequences can be found. Note that harvesting sequences that maximize $J(v^a, v^c)$ for any value of c_1 are termed *Pareto optimal solutions* and the set of Pareto values of J over $c_1 \in [0, 1]$ is termed a *Pareto frontier*.

The problem of finding a Pareto optimal solution for a given initial condition can be a significant computational problem if T is large. Also, the solution obtained is open-loop (nonadaptive) in the sense that it depends only on the initial state of the system (initial means and variance/covariance estimates); that is, it does not make use of any information relating to the estimated current state of the system. Swartzman *et al.* (1987) made use of a feedback algorithm (passive-adaptive in the terminology used by Walters, 1986) that took into account the most recent estimates of the state of the system, and maximized over a set of constraints that reduced the $2T$-dimensional optimization problem to a 2-dimensional optimization problem.[3] The constraints involve knowing the equilibrium solution to the particular Pareto optimization on hand (i.e., the particular choice of c_1). For the case of total U.S. and Canadian yield maximization (i.e., $c_1 = 0.5$ and $c_2 = 0$), this solution will be referred to as the MESY solution (the maximum of the expected sustainable yields associated with criterion (4.47)). The MESY surface is illustrated in Figure 4.14 for Canadian effort levels ranging over 0 to $v^c_{\max} = 20$ units (1 unit of effort equals 2000 standard vessel days) and U.S. effort levels ranging over 0 to $v^a_{\max} = 52$ units of effort.

Two other equilibrium scenarios were analyzed, both corresponding to specific noncooperative solutions of a two-player dynamic game. The first of these is the *min-*

---

[3] The optimal solution to the 2-dimensional problem is only a suboptimal solution to the $2T$-dimensional problem because of the costs associated with the constraints used to reduce the dimension of the problem.

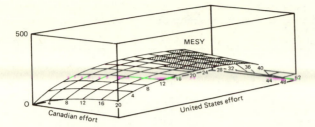

FIGURE 4.14. The expected sustainable yield surface for the combined U.S./Canadian Pacific whiting fishery. The surface is flat around the maximum point (MESY) as indicated by the shaded portion. (From Swartzman *et al.*, 1987.)

max solution. Here each player's strategy is chosen so as to guarantee that at least his min-max yield or revenue value will be obtained (but usually exceeded). The U.S. player, for example, finds his min-max effort level by solving the problem,

$$\max_{v^a \in [0, v^a_{max}]} \min_{v^c \in [0, v^c_{max}]} (Y^a - c_2 v^a),$$

subject to the two-player harvesting model (cf. equations (3.90), (4.45) and (4.46)) in equilibrium. The Canadian player finds his min-max solution in an analogous manner.

If both players assume that the other player is rational, they can take a chance on improving their min-max solutions by playing the *Nash equilibrium solution*. The concept of this solution is explained in the following specific terms. If the U.S. player knew that the Canadian player was going to choose an equilibrium effort level $\hat{v}^c$, then the U.S. player should choose his effort so as to

$$\max_{v^a \in [0, v^a_{max}]} (Y^a - c_2 v^a),$$

subject to the given harvesting model and the given Canadian effort level. If this problem were solved for all Canadian effort levels $v_c \in [0, v^c_{max}]$, we would obtain an optimal

217

TABLE 4.15. Equilibrium strategies for the U.S./Canada two-player Pacific whiting fishery.

Revenue function	Strategy[2]	v^a	v^c	Revenue U.S.[3]	Can.[3]	Total	Y^a	Y^c	Total yield[4]
	Pareto[5]	12	16	156	33	189	156	33	189
Y	Nash	12	16	156	33	156	189	33	189
	min-max	15	16	160	27	160	187	27	187
	Pareto	12	2	164	9	173	176	10	186
$Y - v$	Nash	12	14	146	17	163	158	31	189
	min-max	14	14	147	13	160	161	27	188
	Pareto	8	2	120	6	126	160	16	176
$Y - 5v$	Nash	8	2	120	6	126	160	16	176
	min-max	8	2	120	6	126	160	16	176
	Pareto	4	2	85	1	86	120	28	148
$Y - 10v$	Nash	6	2	85	1	86	145	21	166
	min-max	6	2	85	1	86	145	21	166

[1] For each fishery the revenue function is $Y - c_2 v$ (see equation (4.47)), where the various cases correspond to different choices for the cost/price constant c_2.

[2] All effort units are in 1000 standard boat days.

[3] These are the values obtained for the revenue function in non-dollar units (units based only on relative cost/price values).

[4] Yield units are 1000 metric tons.

[5] This is the Pareto solution when U.S. and Canadian revenue functions are equally weighted ($c_1 = 0.5$ in expression (4.47); i.e., it is the MESY solution for the overall fishery, and for the purposes of comparison with the other strategies, the actual value obtained from expression (4.47) is multiplied by 2.

U.S. effort level $v^{a*}(v^c)$ as a function of v^c. Similarly, the Canadian could generate an optimal effort level $v^{c*}(v^a)$ that was a function of the U.S. effort level v^a. According to Nash, that rational strategy for both players is to play the effort levels that occur at the point where the curves $v^{a*}(v^c)$ and $v^{c*}(v^a)$ overlap (if indeed they do) in the effort space $[0, v^a_{max}] \times [0, v^c_{max}]$. This point is called the *Nash equilibrium*. If a player did not play his Nash equilibrium strategy while his opponent did, then the non-Nash player would do worse (obtain less yield or rent) then he possibly could (which is not rational according to Nash).

The Pareto (MESY), min-max, and Nash solutions for the Pacific whiting fishery are listed in Table 4.15. Note that the results in Table 4.15 are approximate because solutions were found by searching only over integer values of v^a and v^c in the effort space $[0, 52] \times [0, 20]$. As would be expected, the cooperative Pareto strategy always gives the highest total revenue with the lowest effort levels. The rational noncooperative Nash equilibrium gives equal or smaller revenues for the same or greater effort levels while the "not-so-rational" noncooperative min-max gives the lowest total revenue for the highest effort levels. When the cost of effort is not considered ($c_2 = 0$), the Nash and Pareto (MESY) solutions are indicated as being the same. When a unit of yield and a unit of effort have the same relative value ($c_2 = 1$), the MESY Pareto solution can be marginally increased over the Nash solution, but only at a two-thirds reduction in the catch taken by the Canadian fishery. Thus, for this cost structure, it would pay the Canadians to cooperate only if they could bargain a Pareto constant c_1 in expression (4.47) that provided them with a share exceeding seventeen units of revenue.

The dynamic aspects of the problem are much more difficult to analyze in a game setting. Swartzman *et al.* (1987) only analyzed the problem of managing a varying stock in a joint (cooperative) fishery setting where one authority could choose both the U.S. and Canadian effort levels. The quota-determining algorithm they used, as discussed above, was designed to reduce the amount of computational time required to find a "good solution" (constrained optimal solution) to the problem of setting quotas. The algorithm was designed to respond to estimated stock levels by selecting maximum allowable U.S. and/or Canadian effort levels whenever the U.S. and/or Canadian portion of the stock was above its respective MESY stock levels, but only moderately reducing MESY U.S. and/or Canadian effort levels whenever the U.S. and/or Canadian portion

219

of the stock was below its respective MESY stock levels. In particular, Swartzman *et al.* (1987) focused on questions relating to the probability of the stock falling below certain predetermined levels. An alternative, and perhaps better approach, than the above management algorithm might be to solve a parameter optimization problem in which the U.S. and Canadian effort levels are each specified as having a functional form such as equation (4.44).

4.8 REVIEW

We began this chapter by demonstrating the connection between Beverton and Holt's fisheries cohort model and the age-structure matrix models discussed in Chapters 2 and 3. Specifically, the Beverton and Holt cohort model is an age-structured matrix model in which the linear stock-recruitment relationship is replaced by the assumption of constant recruitment. This analysis is only useful under equilibrium conditions (since, at equilibrium, recruitment is constant even though an underlying linear or nonlinear stock-recruitment relationship exists), or if the stock effectively consists of a single cohort, as may happen in some salmon fisheries, or fisheries dominated by a particularly strong cohort recruited in an exceptional year.

We then generalized Beverton and Holt's cohort approach to include a nonlinear stock-recruitment relationship. Thus we essentially showed that (cf. equation (3.26))

$$
\mathbf{x}(t+1) =
\begin{pmatrix}
0 & \cdots & 0 & 0 \\
s_1 & \cdots & 0 & 0 \\
\vdots & \ddots & \vdots & \vdots \\
0 & \cdots & 0 & 0 \\
0 & \cdots & s_{n-1} & 0
\end{pmatrix}
\mathbf{x}(t) +
\begin{pmatrix}
s_0 x_0 \psi(x_0) \\
0 \\
\vdots \\
0 \\
0
\end{pmatrix}
$$

is a suitable fisheries model if we interpret

$$
x_0 = \sum_{i=1}^{n} b_i x_i
$$

as an egg-production index, and assume that the survival parameters s_i include both the natural mortality rates α_i (assumed constant with respect to age i) and the fishing mortality rates $q_i v$ (q_i is a catchability coefficient and v is a constant fishing effort that is applied over the fishing season $[k, k + \bar{t})$). Under this assumption we have

$$s_i = e^{-\left(\alpha_i + q_i v(k) \dot{t}\right)},$$

which leads to our "standard-fisheries" model (assuming $s_n = 0$; cf. (3.34)),

$$x_1(k + 1) = s_0 x_0(k) \psi\big(x_0(k)\big),$$
$$x_{i+1}(k + 1) = e^{-\left(\alpha_i + q_i v(k) \dot{t}\right)} x_i(k), \quad i = 1, \ldots, n - 1.$$

The major difference between this standard-fisheries model and a "complete-harvesting" model such as (cf. (4.21))

$$x_1(k + 1) = s_0 x_0(k) \psi\big(x_0(k)\big),$$
$$x_{i+1}(k + 1) = e^{-\alpha_i} x_i(k) - e^{-\alpha_i(1 - t_i^\circ)} u_i, \quad i = 1, \ldots, n - 1,$$

is that each age class cannot be individually harvested in the standard-fisheries model. The fact that individuals in the ith age class in the complete-harvesting model are harvested only at time t_i° is not important since harvesting can always be modeled as occurring over a time interval, say $[k, k + \bar{t})$, by selecting a harvest rate v_i and replacing the right-hand side $e^{-\alpha_i} x_i(k) - e^{-\alpha_i(1 - t_i^\circ)} u_i$ in the complete-harvesting model with $e^{-(\alpha_i + q_i v_i(k)\dot{t})} x_i(k)$; that is, each age class experiences a different v_i, $i = 1, \ldots, n$.

We referred to the maximum sustainable yield solution associated with the complete-harvesting model as the USY (ultimate sustainable yield) since the inability to harvest all age classes independently in the standard-fisheries model implies that the MSY (maximimum sustainable yield) associated with the standard-fisheries model is always less than or equal to the USY. Although the USY is not attainable in most fisheries, it is still useful to calculate its value since, as discussed with reference to the South African anchovy fishery (see Figure 4.3), the USY provides a point of reference for how far the current fishing policies fall short of the yield potential of the resource. Under assumptions such as constant mortality with age, and fecundity proportional to size, we demonstrated in Chapter 3 how an age-class model can be reduced to a lumped single-variable model. Single-variable models, however, cannot be used to calculate such age-related entities as the USY. Thus even if a single-variable model is useful for describing the current operation of a fishery, its application is limited in exploring a number of age-related regulatory questions (e.g., deciding on appropriate gear), or evaluating the true potential of the resource to provide yield under improved technologies. Again, from an age-structured analysis of the South African anchovy fishery (Figure 4.3), it is apparent that a larger mesh than that used at the time of the analysis (Getz, 1980a) would probably lead to increased biomass yield.

Equilibrium analyses are limited because fisheries are subject to stochastic environmental influences. Further, they do not provide us with information on how to explore virgin or overexploited stocks. Whether stochastic or dynamically deterministic, from the management case studies presented in Sections 4.4 and 4.5 it is apparent that yield, fishing-effort, and stock levels cannot be simultaneously "stabilized." For example, a constant-catch policy requires that we allow the stock to fluctuate and, in the long run, if a constant-catch policy does not respond to severe weaknesses in the stock, it can destroy the fishery altogether. On the other hand, a fixed-escapement policy is designed to stabilize the stock, but only at the expense of the catch, possibly even closing down the fishery in years when the stock is particularly weak. Finally, a constant-effort policy has the advantage of stabilizing labor and capital invested in the fisheries operation, but only at the expense of transferring variability from the stock to the catch. Results from the various case studies, however, indicate that constant-effort policies exhibit less yield variability than fixed-escapement policies, and greater stock stability than constant-catch policies. If we are able to assign cost to variability associated with fluctuations in catch, effort, and stock policy respectively, around some appropriate MSY levels Y^*, v^*, and $\mathbf{x}^*$—that is, assign relative values to the parameters π_j, $j = 1, 2, 3$, in a criterion such as (cf. expressions (4.25) and (4.26) and recall that w_i is the weight of an individual in age class i)

$$
\begin{aligned}
J_T(\mathbf{x}_0, \mathbf{x}(T)) = {} & \pi_1 \sum_{k=0}^{T-1} (Y_k(v(k), \mathbf{q}) - Y^*)^2 \\
& + \pi_2 \sum_{k=0}^{T-1} (v(k) - v^*)^2 \\
& + \pi_3 \sum_{k=1}^{T} \sum_{i=1}^{n} (w_i x_i(k) - w_i x_i^*)^2,
\end{aligned}
$$

for example—then the Monte Carlo simulation techniques can be used to estimate the optimal set of parameters, $\tilde{q}_1$, $\tilde{q}_2$, $\tilde{x}_0$, and ρ, for a quota-setting policy (i.e., selecting Y_k) of the form (see Section 4.5.4)

$$
Y_k = \left(\frac{\tilde{q}_1 \left(x_0(k) - \tilde{x}_0 \right)}{\tilde{q}_2 + \left(x_0(k) - \tilde{x}_0 \right)} \right)^\rho .
$$

In the final section of this chapter we showed how the standard-fishery model is easily extended to a technology-linked multispecies setting, to account for mixed-species catches or landings at the dock. Of course, several lumped single-variable models can be combined to obtain a less elaborate technology-linked multispecies model than equations (4.37). However, Pikitch's (1987) analysis of the U.S. West Coast groundfish resource provides an excellent example of how age-structured models can be used to determine optimal gear size in multispecies fisheries. Further, the question of transboundary migration as a function of age in the U.S.-Canadian Pacific whiting fishery could not be adequately dealt with if age structure were not accounted for in the multiparticipant model based on equations (4.45) and (4.46).

CHAPTER FIVE

Forest Management

5.1 BACKGROUND

The problem of determining sound timber harvest policies traditionally has been separated into stand-level and forest-level problems. A timber stand is an administrative unit of the forest that is bounded by physical and geographic features such as roads and drainages and by discontinuities in the forest vegetation. The stand-level problem involves determining the sequence of management activities including harvesting and planting for an existing stand. A forest ownership is a collection of many stands, and the forest-level problem involves determining the management intensity and flow of timber products from the forest.

All but the last two sections in this chapter are concerned with stand-level problems. To evaluate alternative harvest systems, stand growth in response to harvesting and other silvicultural treatments must be predicted. In Section 5.2 we review univariate models and present linear and nonlinear stage-class models for the prediction of stand growth and yield. A nonlinear stage-class model for mixed white-fir and red-fir stands in California is developed, and its predictions are compared with those obtained from a more complicated single-tree simulator.

Resource managers define silvicultural policy for existing and future stands by selecting one of two systems for long-term timber production: even-aged and uneven-aged management. The debate over which system should be applied often centers around economic efficiency in addition to the impacts of the two systems on other resources such as wildlife habitat or visual quality. In Section 5.3

we present a model for evaluating and comparing the efficiency of the two harvest systems. The model is a generalization of the traditional Faustmann formula for determining the optimal rotation age for an even-aged plantation, and it includes a stage-class model for projecting growth and yield in even-aged and uneven-aged stands. Even-aged and uneven-aged harvesting problems are discussed in detail, and measures of economic efficiency are compared.

The size and complexity of harvesting problems that include stage-class models for predicting growth and yield has led to the use of numerical solution methods for computing optimal solutions. In Section 5.4 we describe how to solve even-aged and uneven-aged stand harvesting problems with gradient methods and penalty functions.

In Section 5.5 we present two case studies in which numerical techniques are used to solve stand-level harvest problems. First, finite planning horizon approximations to an infinite time horizon problem are solved to determine optimal harvest strategies that convert existing uneven-aged stands of California white fir and red fir to desired steady-state stand structures. Results are compared for two approximation methods under various initial conditions, lengths of planning horizons, and discount rates. Second, optimal even-aged and uneven-aged management regimes are determined for young-growth and old-growth true fir stands, and the relative efficiency of the two harvesting systems is compared.

In Section 5.6 we describe the application of stage-class models to forest-level problems of determining optimal harvest levels from large timber ownerships. In the final section (5.7), we review the application of stage-class models in forestry management.

226

5.2 MODELS FOR PREDICTING STAND GROWTH

5.2.1 Stand Management Systems

In contrast to fisheries, measuring and harvesting forest stands is relatively easy and has led to a variety of stand management systems and associated models for predicting stand growth and determining harvest policy. A stand management system is a planned program of silvicultural treatments during the whole life of a stand. Silvicultural treatments include regeneration cuttings that are used to establish a new stand by means of natural or artificial regeneration. Intermediate cuttings such as thinnings or selection harvests are used to control the density and growth of existing stands.

Stand management systems are most often classified on the basis of their reproduction method (see, for example, Smith, 1962). The most well-known management system uses the clearcutting reproduction method and produces even-aged stands. In this case, even-aged management is characterized by a cycle of events (rotation period) that includes clearcutting a mature stand, planting a single cohort of trees, periodically thinning the new crop, and clearcutting the crop at a specified rotation age. Even-aged management is also conducted using seed-tree and shelterwood regeneration methods. With the seed-tree method, most of the mature trees are removed in one cut except for a small number of trees that are left to provide seed for the next stand. After the new crop of trees has become established, the seed trees are cut. The shelterwood method involves the removal of the mature trees in a series of cuttings that extend over a relatively short period of time. In these latter two methods, natural regeneration is established under the partial shelter of the seed trees.

Uneven-aged management uses a selection harvesting method that involves the periodic harvest of trees in spec-

227

ified size classes. Selection harvests are conducted to control the spacing and growth of the remaining trees and to enhance natural regeneration from seeds produced by the remaining mature trees. Regeneration in uneven-aged stands may also be done with planting. Since selection harvesting and regeneration take place simultaneously, uneven-aged stands include a mixture of trees in a wide range of size and age classes.

5.2.2 Univariate Models for Even-aged Stands

The production model for even-aged management expresses stand volume or value as a function of stand age; the shape of the relationship can be characterized by a saturating growth curve (Figure 5.1) (for example, see Clark, 1976). Management analyses are conducted using this growth curve as the basis for a general lumped parameter model; and this model is used to examine the conditions for the optimal rotation age and planting intensity using the classic Faustmann (1849) formula (for example, see Clark, 1976). The general univariate model has been used to determine the optimal rotation period for various management objectives (Chang, 1984) and changes to the economic assumptions in the Faustmann formula (see Reed, 1986 for review). The univariate model has been extended to include stand volume as an independent variable so that optimal thinning patterns and rotation ages can be determined simultaneously (Clark, 1976; Cawrse *et al.*, 1984).

The univariate model is realistic when stands contain trees with uniform size or when economic value does not depend on the distribution of tree sizes. However, the model lacks the biological detail needed for projecting stand attributes in cases where economic value at rotation age depends on the distribution of trees by size class. In these cases the continuous time univariate model has been replaced by discrete-time *whole-stand* models. Whole-

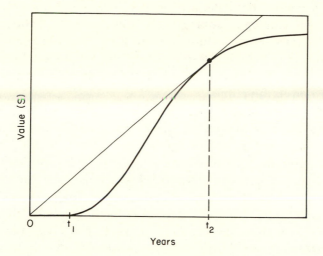

FIGURE 5.1. The value of a stand as a function of the age of the stand since it was last clearcut. The stand has no value until age t_1. The value of the stand divided by age (average value accumulated per unit time) is maximized at t_2. (For more details see Clark, 1976.)

stand models include equations that predict future stand density (e.g., volume, basal area, number of trees) as a function of the current stand age and density, and they include equations that predict the parameters of a probability density function for the distribution of trees by diameter class as functions of the aggregate stand attributes (see, for example, Knoebel *et al.*, 1986).

The estimation of the parameters in this system are made easy by the detailed growth observations in stand inventories. Stand inventory and growth observations consist of a set of permanent plots that are remeasured periodically, or they consist of a set of temporary plots in which past growth is measured using increment cores or stem profiles from the existing trees. Growth can be measured in temporary plots because tree stems contain rings that act as records of past growth. In permanent and tem-

229

porary plots, both the attributes of individual trees (such as stem diameter and height) and the attributes of the plot (such as total volume and number of trees) can be tracked. Thus, the observations of the distributions of trees by size class over time can be used to estimate the parameters of the density functions.

While whole-stand models give more detailed descriptions of stand structure, the shapes of the distribution functions and the use of age as a prediction variable limit the application of whole-stand models to the projection of even-aged plantations or natural stands. In stands that include trees of many ages and species, either even-aged or uneven-aged management may be employed. In contrast to whole-stand models, stage-class models contain explicit equations for the growth of trees between size classes and for regeneration. Thus, stage-class models provide a more flexible framework that allows the projection of growth and yield in both even-aged and uneven-aged stands. Stage-class models are systems of discrete-time difference equations, and in the next two subsections we describe both linear and nonlinear forms.

5.2.3 Linear Stage-Class Models

In Chapters 2 and 3, we developed a compact notation for the discrete-time stage-structured harvesting model (cf. equation (3.45)):

$$\mathbf{x}(t + 1) = G\mathbf{x}(t) + \boldsymbol{\phi} - \mathbf{u}(t), \tag{5.1}$$

where G is a stage-progression matrix, $\boldsymbol{\phi}$ is an input vector (which in the nonlinear model depends on the state of the population), and $\mathbf{u}$ is a vector of harvest variables. Applied to a forest stand, each element of the state vector $\mathbf{x}(t) = (x_1(t), \ldots, x_n(t))'$ represents the number of trees in a specified growth stage at the beginning of time period t. Stages are defined by tree diameter (at breast height)

classes. Tree diameter is a convenient attribute for defining growth stages because it is easy to measure and it is
related to tree volume and other attributes that define the
value of the tree. The elements $u_i(t)$ of the control vector $\mathbf{u}(t)$ represent the number of trees harvested from the
ith stage just prior to transition (see Figure 2.1), and the
dimension of $\mathbf{u}(t)$ is equal to n.

The matrix $G = PS$ (cf. (2.24)–(2.26)) represents the
progression and mortality processes of the tree population. Recall from equation (2.26) that G has the form

$$G = \begin{pmatrix} (1 - p_1)s_1 & \cdots & 0 & 0 \\ p_1 s_1 & \cdots & 0 & 0 \\ \vdots & \ddots & \vdots & \vdots \\ 0 & \cdots & (1 - p_{n-1})s_{n-1} & 0 \\ 0 & \cdots & p_{n-1}s_{n-1} & s_n \end{pmatrix},$$

where the parameters s_i and p_i take values on the interval $[0, 1]$, and respectively denote the proportion of trees
in stage i at time t that survive the time period $(t, t + 1]$
and move into the next growth stage at the end of time
period t. We assume that the time interval and diameter class widths are defined so that trees advance at most
one stage in each time period. To satisfy this assumption,
we use a 5-year growth interval and 2-inch diameter in
the stage-class model for true fir described later in this
section. If trees advance more than one stage during a
projection interval, then additional off-diagonal elements
can be added to the stage-progression matrix G (see, for
example, Hann, 1980). When trees are allowed to advance more than one growth stage in a projection interval,
smaller diameter class widths may be used to characterize
tree size distributions.

In this chapter we make the natural assumption that
only the first element ϕ_1, of the input vector ϕ, is nonzero
(i.e., trees don't migrate). In forest modeling the function
ϕ_1 is commonly termed the *ingrowth function* (in Chapters 3

231

and 4 we referred to it as the stock-recruitment function), since it represents the number of saplings that germinate from seed on the forest floor. In this section, we consider the case where ingrowth is linearly dependent on the number of trees in each stage class. At least three kinds of linear ingrowth functions have been formulated. Ingrowth may be determined by total seed production. For example, in a stage-class model for California redwoods, Bosch (1971) defined ingrowth as $\phi_1 = s_0 \sum_{i=1}^n b_i x_i(t)$, where s_0 is the seedling survival rate and b_i is the number of seeds produced per tree in stage class i (see also the discussion of the stage-class model, equation (2.27) in Chapter 2). This ingrowth function may be appropriate for a relatively young stand where space is often available for seeds to germinate. In older stands, where seedling establishment is only possible in spaces provided by the removal of existing trees, ingrowth depends on the numbers and stages of trees removed in the previous harvest, that is, $\phi_1 = \sum_{i=1}^n b_i u_i(t-1)$, where b_i is the number of new recruits per tree harvested in stage i (see, for example, Usher, 1966). Buongiorno and Michie (1980) combined the concepts of seed production and crowding by formulating an ingrowth function in which seedling establishment increases linearly with seed production, but decreases linearly as a function of the basal area of the stand (stand basal area is the total cross-sectional area of tree stems measured at breast height and is related to the amount of open space available for seedling establishment). Specifically, in each time period, Buongiorno and Michie (1980) assumed that the ingrowth function is given by

$$\phi_1 = \beta_0 + \beta_1 \sum_{i=1}^n x_i(t) - \beta_2 \sum_{i=1}^n b_i x_i(t), \qquad (5.2)$$

where $\beta_0 \geq 0$ is the ingrowth independent of stand structure (e.g., seeds borne on the wind), $\beta_1 > 0$ is the average number of seeds produced per tree, $\beta_2 > 0$ is the mortality factor, and b_i is the basal area of a tree in stage i. A more general function of this type would allow for a variable seed production for trees in different stage classes, but as discussed in the next section, the linear approach has limited application.

Stand management applications with linear stage-class models have centered around the determination of maximum sustainable yield harvest policies for uneven-aged stands. While early applications defined steady-state yield as a single harvest variable representing the proportion of trees removed across all tree size classes (Usher, 1969a,b), later studies allowed the determination of size-specific harvest rates (e.g., Rorres, 1978).

The maximum sustainable yield problem with size-specific harvest rates is defined as follows. A steady state exists if $\mathbf{x}(t + 1) = \mathbf{x}(t) = \mathbf{x}$, say, for $t = 1, 2, \dots$. Further, let $\mathbf{w} \in R^n$ be the vector of volume yields from trees in various growth stages. With these definitions, the maximum sustainable yield problem is (cf. equation (2.40) and Problem 1 in Section 2.5.2):

$$\max_{\substack{\mathbf{x} \geq 0 \\ \mathbf{u} \geq 0}} J = \mathbf{w}'\mathbf{u}, \tag{5.3}$$

subject to the steady-state constraint

$$\mathbf{u} = (G - I)\mathbf{x} + \boldsymbol{\phi}, \tag{5.4}$$

which is obtained by reorganizing the growth dynamics (5.1) subject to the steady-state condition. This maximum sustainable yield problem is easily solved using linear programming.

When stand growth is defined with a linear stage-class model, the form of the recruitment model ϕ_1 determines whether or not the maximum sustainable yield problem (5.3) is bounded. Suppose that recruitment is directly

233

proportional to the stand state or harvest level, and suppose that a feasible solution $(\mathbf{x}, \mathbf{u})$ to the steady-state constraint (5.4) can be obtained with yield $\mathbf{w'u} > 0$. Then, for any scalar $\alpha > 0$, the pair $(\alpha\mathbf{x}, \alpha\mathbf{u})$ is also a feasible solution to the steady-state constraint, and as α increases without bound, the steady-state yield $\alpha\mathbf{w'u}$ increases without bound. This is consistent with the result, obtained in Chapter 2, that the solution to the maximum sustainable yield problem depends linearly on the size of the population (see equation (2.48)). To bound the problem, a growing stock constraint must be used in addition to the steady-state constraint (5.4). Then the maximum sustainable yield solution can be found for some specified stocking level (for example, see Rorres, 1978).

If we define ingrowth as a proportion of the stand state or harvest control, then two problems arise in the context of determining maximum sustainable yield harvest policies. First, determining the optimal harvest policy for a given stand density leaves the problem of determining the optimal density unsolved. Second, defining ingrowth to be proportional to the stand state ignores the observation that ingrowth decreases as stand density gets large. Defining ingrowth as a constant modified by both positive and negative effects of stand density (cf. equation (5.2)) helps to avoid these problems. Steady-state ingrowth depends on the constant β_0 and decreases as the stand basal area increases. In the maximum sustainable yield problem, this basal area feedback bounds the optimal steady-state growing stock without an explicit growing stock constraint (see, for example, Buongiorno and Michie, 1980).

The estimation of the growth and survival parameters in G (equation (5.1)) and the parameters of the ingrowth function ϕ_1 depends on the kind of data available (see Michie and Buongiorno, 1984, for a detailed discussion of growth parameter estimation). A data set that includes repeated observations of tree diameters in permanent plots that are located in a variety of different stands contains the most information. The data for each plot include the number of trees in each stage that either remain in the same stage, move up one stage, or are lost to mortality or harvest between two successive measurements. Unbiased estimates of growth and survival parameters can be obtained by computing p_i and s_i, $i = 1, \ldots, n$ for each plot remeasurement and then computing the means for p_i and s_i, $i = 1, \ldots, n$ across all plots and time periods. Unbiased estimates of the parameters for the ingrowth equation can be obtained using ordinary least squares by treating the plot density for each remeasurement as an observation.

A second type of data set includes repeated observations of the diameter distributions and number of trees harvested in permanent sample plots. Data sets of this type are cheaper to obtain because individual tree growth is not recorded. To estimate the model parameters, the growth dynamics on the right-hand side of (5.1) are collapsed into the matrix A, as in equation (2.22), to obtain

$$\mathbf{x}(t + 1) = A\mathbf{x}(t) - \mathbf{u}(t). \tag{5.5}$$

The diagonal and off-diagonal elements in A, which represent, respectively, the proportions of trees that survive and stay in the same stage and the proportions of trees that survive and advance one stage, can be estimated simultaneously using generalized least squares, or the elements in each row of A can be estimated independently using ordinary least squares. In both cases, the parameters may be biased because the errors in the growth equations for adjacent stages are likely to be correlated and because, with repeated observations, the lagged dependent variable $\mathbf{x}(t)$ appears as an independent variable on the right-hand side of equation (5.5). To avoid these problems, Michie and Buongiorno (1984) propose a third estimation method. Extraneous guesses of mortality rates can be substituted into the elements of the growth matrix A (see expression (2.22)) so that the growth parameters in each row can be estimated sequentially using ordinary least-squares starting with row n working backwards to row 1. In each row, only one parameter is estimated because of the substitution of the parameter estimate from the previous row.

Michie and Buongiorno (1984) analyzed a data set that contained repeated observations of tree diameters in plots located in hardwood stands in Wisconsin. They computed the unbiased mean proportions for growth and survival directly from the data. They compared these estimates with those obtained from least-squares estimation using equation (5.5) and the same data set, and they concluded that the least squares estimates were seriously biased. They also estimated the growth parameters by inserting extraneous mortality estimates into the matrix A in equation (5.5). The resulting growth parameter estimates were unbiased relative to the proportions computed using the full data set. These results show that growth estimates can be obtained without having repeated growth observations of individual tree diameters.

The attraction of the linear stage-class model is its simplicity. Each parameter of the model has a direct interpretation, and the parameters can be estimated with ordinary linear regression or in some cases with simple multiplications and divisions. Future stand states and the impact of harvesting can be determined analytically. Finally, optimal harvesting regimes can be determined using standard linear programming packages. The tractablility of the linear stage-class model is gained with the major assumption that tree growth and mortality rates are fixed and independent of stand density. However, many studies based on repeated observations of tree growth and mortality across plots with different densities have concluded that plot density does have a significant impact (for example, see Cole and Stage, 1972, and Hamilton, 1986).

To take into account the impacts of stand density on tree growth, survival, and ingrowth, stage-class models can be formulated such that the elements of the growth matrix G vary with stand density. In the next three subsections we discuss applications of density-dependent nonlinear stage-class models in forestry.

5.2.4 Nonlinear Stage-Class Models

In this subsection we present a nonlinear stage-class model that includes the linear stage-class model discussed above as a special case. The nonlinear model includes the growth matrix $G = PS$ and the regeneration vector ϕ; however, in contrast to the linear model, the elements of P, S, and ϕ are nonlinear functions of the stand state x. The nonlinear relationships for tree growth, survival, and ingrowth give more realistic portrayals of stand growth processes, and thus we apply the nonlinear stage-class model to forest management questions later in this chapter.

In a nonlinear stage-class model for stand growth, the stage-class transition and survival rates and the ingrowth levels are nonlinear functions of various measures of stand density, which are aggregations of the elements x_i in the state vector x. A stage-class model may use m density measures represented by the m-dimensional vector y, which is computed as a linear transformation of the state vector x. Recalling equation (3.46) in which Θ is an $m \times n$ aggregation matrix containing parameters θ_{ij}, the vector of stand density measures is obtained with the linear transformation

$$\mathbf{y}(t) = \Theta\mathbf{x}(t). \tag{5.6}$$

The most commonly used density measures are the stand basal area (the total cross-sectional area of the tree stems measured 4.5 feet from the ground), the number of trees, and the total cross-sectional crown area of the stand. For example, if y_1 is defined as the total stand basal area, the first row of the matrix Θ contains elements θ_{1j} representing the average basal area per tree in size class j, $j = 1, \ldots n$. We now assume that the parameters p_i, s_i, and ϕ_i depend on $\mathbf{y}$.

Just as in the linear stage-class model (5.1), each element of the state vector $\mathbf{x}(t)$ represents the residual number of trees in a specified growth stage at the beginning of time period t where the growth stages are defined by tree diameter (at breast height) classes. The matrix G now contains elements $g_{ij}(\mathbf{y})$, which are products of the scalar functions for the survival and transition rates $s_i(\mathbf{y})$ and $p_i(\mathbf{y})$ (see equation (3.47)). These functions take values on the interval $[0, 1]$ and denote the proportion of trees in stage i at time t that, respectively, survive the time period $(t, t+1]$ and move into the next growth stage at the end of time period t. Stage-class transition rates are decreasing functions of the stand density (e.g., Adams and Ek, 1974). Tree survival rates may be decreasing functions of stand density (e.g., Hamilton, 1986), or they may be fixed parameters that depend only on the diameter class associated with each growth stage (e.g., Adams and Ek, 1974).

The input vector $\phi(\mathbf{y}(t))$ now contains the scalar function $\phi_1(\mathbf{y}(t))$ representing the number of trees entering the smallest growth stage through natural regeneration processes during time period t, and $\phi_i = 0$, $i = 2, \ldots, n$. Early stage-class models used ingrowth equations to predict the number of trees growing into the smallest merchantable diameter class (e.g., Adams and Ek, 1974). Ingrowth was inversely related to stand basal area, and for a given basal area was directly related to the number of trees. These ingrowth models assumed that trees in each stage produced the same number of seed. Later models represented natural regeneration as the product of the number of seeds produced and the seedling establishment rate (for example, see Haight, 1987). Seed production depended on tree size, seedling establishment was a decreasing function of stand density, and the state vector $\mathbf{x}$ included seedling and sapling growth stages for trees less than 4.5 feet tall.

Just as in the linear stage-class model, the elements $u_i(t)$ of the control vector $\mathbf{u}(t)$ represent the number of trees harvested from the ith stage just prior to transition (see Figure 2.1), and the dimension of $\mathbf{u}(t)$ is equal to n. For generality we allow an immediate harvest $\mathbf{u}(0)$ from a stand with initial state $\mathbf{x}(0) = \mathbf{x}_0$ before the dynamics begin. The stand growth dynamics with the nonlinear stage-class model are now

$$\mathbf{x}(1) = \mathbf{x}(0) - \mathbf{u}(0)$$
$$\mathbf{x}(t+1) = G(\mathbf{y}(t))\mathbf{x}(t) + \phi(\mathbf{y}(t)) - \mathbf{u}(t),$$
$$t = 1, 2, \ldots. \tag{5.7}$$

Nonlinear stage-class models are constructed by estimating the parameters of each component equation separately. Then the equations are inserted in the matrix framework to project stand growth. The component equations are estimated from repeated observations of tree sizes, mortality, and ingrowth in permanent inventory plots located in stands with different densities and structures. Nonlinear stage-class models have been constructed for northern hardwood stands in Wisconsin (Ek, 1974) and Ponderosa pine in Arizona (Hann, 1980).

5.2.5   Single-Tree Simulators

A second and more complicated method for projecting multispecies uneven-aged stands involves the use of single-tree simulators. In contrast to stage-class models in which the stand state is described with a diameter distribution, single-tree simulators describe the stand with a list of tree records. Each tree record contains the current tree dimensions such as stem diameter, height, and crown length. Single-tree simulators include equations that predict the change in the tree dimensions for each tree record during discrete-time intervals. All single-tree simulators include equations for diameter growth, and many include equations for height growth and tree crown development. The growth equations are functions of either the location of the tree relative to its neighbors (distance-dependent models) or aggregate stand density variables (distance-independent models). In the former case, each tree is described by its spatial coordinates which change due to harvesting or mortality. This methodology has not been used widely because of the time required to obtain tree coordinates. In the latter case each tree record includes a tree expansion factor that represents the number of trees of its kind in the stand. The tree factor is reduced due to harvesting or mortality, and aggregate stand attributes are obtained by summing the products of tree attributes and tree factors over all tree records. Generally, single-tree simulators allow for the projection of 500 tree records or more.

Like stage-class model construction, single-tree simulators are constructed by first estimating the tree growth equations separately and then inserting them in a simulation shell for projecting stand growth. Major modeling efforts in different regions of the U.S. have used the single-tree distance-independent simulation paradigm. These include the STEM simulator for mixed hardwood and conifer stands in the North-Central region (Belcher *et al.*, 1982), the Prognosis model for mixed conifer stands in the northern Rocky Mountains (Wykoff *et al.*, 1982), the SPS simulator for Douglas fir in the Pacific Northwest (Arney, 1985), and the CACTOS simulator for mixed conifer stands in California (Wensel and Koehler, 1985).

There are two key assumptions in the stage-class model that differentiate it from a single-tree simulator. First, in the stage-class model, trees are classified by stem diameter.

Trees are assumed to be uniformly distributed across each diameter class, and a proportion of the trees in each stage are projected to grow into the next stage using a density-dependent diameter growth model. Second, each diameter class represents a growth stage that is described by average tree height, volume, and crown attributes. These class attributes are assumed to be fixed for the course of the projection. By relaxing these two assumptions, the stage-class model takes the form of a single-tree simulator.

An important property of single-tree simulators is the detail in which a stand and its growth processes can be described. Since single-tree simulators explicity model the change in tree diameter, height, and crown dimensions, the impacts of stand density competition and harvesting on these dimensions can be predicted. Further, the number of tree records that can be used to describe the stand is very large. However, this detail is gained at the expense of tractability for management analysis: the number and relationship of equations included in a single-tree simulator has generally precluded the determination of optimal harvesting behavior. Further, the number of interrelated equations in a typical simulator may cause the propagation of substantial projection errors (Gertner and Dzialowy, 1984).

It is our thesis that the stage-class modeling framework provides a balance between the detail required to make meaningful stand projections and the tractability needed to conduct management analysis. In the next two subsections we construct a model for uneven-aged stands that contain mixtures of California white fir and red fir, and we contrast the projections of the stage-class model with the projections from a single-tree simulator for the same stands. The results show that the stage-class model predicts stand development with the same detail, and the results suggest that both models will project stand attributes with the same degree of accuracy when compared to actual stand growth observations.

5.2.6 A Stage Class Model for True Fir

In this subsection we describe work by Haight and Getz (1987a) and construct a stage-class model (STAGE) for white-fir and red-fir stands in Northern California. The model is centered around the diameter growth equations for true fir obtained from the California Conifer Timber Output Simulator (CACTOS), a single-tree simulator for mixed conifer stands (Wensel and Koehler, 1985). While CACTOS includes growth equations for six conifer species, we constructed STAGE to project white fir and red fir because these species grow in pure and mixed stands and because a considerable portion of the timber harvest in California comes from these kinds of stands.

In this section we extend the basic form of G and ϕ in equation (5.7) to represent a two-species model. The stage-structured model is constructed by first defining growth stages for each species. The state vector $\mathbf{x}$ in equation (5.7) includes forty-two stages that are divided into two sets $(x_1, \ldots, x_{21})$ and $(x_{22}, \ldots, x_{42})$ representing the number of trees in white-fir and red-fir growth stages, respectively. (The explicit form of G associated with this model is given in Haight and Getz, 1987a.) For management analysis, trees are separated into sapling, pole, and sawtimber stages. The sapling stages, (x_1, x_2, x_3) for white fir and (x_{22}, x_{23}, x_{24}) for red fir, represent the number of trees in two-foot height classes that range between zero and six feet. The pole and sawtimber stages, $(x_4, \ldots, x_{21})$ for white fir and $(x_{25}, \ldots, x_{42})$ for red fir, represent the number of trees in two-inch diameter (at breast height) classes. Trees in the ith stage are assigned the midpoint diameter d_i, where $d_4 = 1, \ldots, d_{21} = 35$ for white-fir, and $d_{25} = 1, \ldots, d_{42} = 35$ for red fir. Two-inch diameter classes were chosen because the potential diameter growth of white-fir and red-fir trees growing in stands with moderate site quality is less than two inches in five years (Wensel and Koehler, 1985). The maximum tree size is limited by the data used to construct the white-fir and red-fir growth equations in CACTOS (see Biging, 1984).[1]

Tree diameter is a convenient attribute for defining growth stages because it can be related to tree height, volume, and crown attributes. The average tree height was computed for each stage using equations described by Van Deusen and Biging (1985). Average tree volume was computed with the assumption that trees in each stage are sectioned into 16.5-foot logs starting at a one-foot stump height. The top diameter and volume of each log were computed with equations given by Biging (1984). Logs that have top diameters less than six inches were assumed to be unmerchantable. Trees less than seven inches in diameter (at breast height) were assumed to be less than 17.5 feet tall and thus did not contain any logs. Average tree volumes were computed by summing log volumes. The average volume of trees in each red-fir growth stage differed by less than 10% from the average volume of trees in the corresponding white-fir stage.

The models for diameter growth and regeneration depend on measures of the cross-sectional crown area of the stand. The ith element of the stand density vector $\mathbf{y}$ is defined as the cross-sectional crown area of the stand measured at 66% of the height of trees in stage i. Thus, Θ in equation (5.6) is a 42×42 matrix, and the ijth element of Θ represents the cross-sectional area of trees in stage j measured at 66% of the height of trees in stage i. The elements of the jth column of Θ represent a profile of the

---

[1] In the stage-structured model constructed here, CACTOS diameter growth equations are applied to trees between one and five inches in diameter. The minimum tree size in the data set used to construct the diameter growth equations was six inches (Biging, 1984). Thus, the projections of pole-size trees should be viewed with caution.

243

cross-sectional crown areas of a tree in stage j. These elements were calculated using equations for crown volume, height to crown base, and crown area given by Wensel and Koehler (1985). Compared to the crown profile of a white-fir tree in a given diameter class, the crown areas of a red-fir tree in the same diameter class are smaller at the base of the crown and larger near the top of the crown.

The structure of the matrix G in expression (5.7) is modified from the structure defined by equation (3.47) to allow the projection of the two-species state vector $\mathbf{x}$. Transition equations $p_1, \ldots, p_{21}$ and $p_{22}, \ldots, p_{42}$ are defined for the white-fir and red-fir growth stages, respectively. Trees are projected to grow between pole and sawtimber stages using diameter growth models given by Wensel and Koehler (1985). Diameter class transition rates are the products

$$p_i(\mathbf{y}) = \varphi_i \chi_i(\mathbf{y}), \qquad i = 4, \ldots, 21, 25, \ldots, 42,$$

where φ_i is the maximum proportion of trees in stage i that move up one diameter class under no competition, and $\chi_i(\mathbf{y})$ is the proportion of the maximum amount of growth that can be achieved with a given level of competition. The potential diameter class transition rates are time invariant and depend on the productive capability of the site. In our simulations, we assume that the potential growth rates are the same for each species and represent the growth potential for trees on moderately productive sites (see Wensel and Koehler, 1985). Finally, trees in each species are not allowed to grow beyond the 35-inch growth stage so that $p_{21} = 0.0$ and $p_{42} = 0.0$.

The impact of density-dependent competition on potential white-fir and red-fir diameter class transition rates is

$$\chi_i(\mathbf{y}) = \begin{cases} e^{-2.544\left(\frac{y_i}{43560}\right)^{0.8458}}, & \text{if } i = 4, \ldots, 21; \\ e^{-10.500\left(\frac{y_i}{43560}\right)^{0.1941} c_i^{-0.28}}, & \text{if } i = 25, \ldots, 42, \end{cases} \quad (5.8)$$

where y_i is the stand crown cover measured at 66% of the height of trees in stage i, and c_i is the average crown volume for trees in stage i. The competition functions for

white fir depend on stand density while the functions for
red fir depend on crown volume in addition to stand density. Red-fir crown volumes were computed using equations from Wensel and Koehler (1985). All levels of crown
cover produce greater reductions in the growth rates of
red fir with diameters less than seven inches compared to
the impacts of crown cover on white-fir diameter growth
(Figure 5.2). Conversely, crown covers greater than 40%
produce smaller reductions in the growth rates of red fir
greater than 15 inches in diameter than do the same levels
of crown cover on white-fir diameter growth.

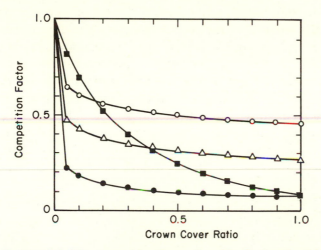

FIGURE 5.2. The competition factors for white fir (■) and red fir
with diameters (at breast height) equal to 29 inches (o), 15 inches
(△), and 7 inches (●) are plotted against crown cover percentages.

The data used by Wensel and Koehler (1985) to construct the diameter growth models, as discussed by Biging
(1984), came from thirty-eight plot clusters located in the
mixed conifer region of Northern California. The diameters, heights, and crown lengths of all trees on each plot

were recorded. These were used to compute the competition levels. Diameter growth was measured by felling selected trees and cutting them into sections to record their diameter growth over time and by taking diameter increment cores from the standing trees.

There is some question about the biologically correct shape for the competition models. The negative exponential functions chosen by Wensel and Koehler (1985) result in immediate reductions in diameter growth for small levels of crown cover in the stand. Alternatively, intertree competition may take place only after a significant amount of crown cover is present, perhaps as much as 30%. In this case, the competition models would be sigmoid and asymptotic to the upper axis. It turns out that the shapes of the competition models have a profound impact on the species composition of optimal harvesting regimes. Because of the uncertainty in the form of the competition model, the structures and species compositions of optimal harvest regimes presented in Section 5.5 should be viewed accordingly. Alternative forms for modeling diameter growth can be found in the literature describing single-tree simulators (Belcher *et al.*, 1982; Wykoff *et al.*, 1982; Arney, 1985).

Survival equations based on observations of white-fir and red-fir survival rates were not available. Instead, the 5-year survival rates for poles and sawtimber were projected with the function,

$$s_i = 1.0 - 0.1e^{-0.1d_i} \qquad i = 4, \ldots, 21, 25, \ldots, 42, \quad (5.9)$$

which was adapted from a survival equation for Douglas fir in the North Coast region of California given by Wensel and Koehler (1985). Survival rates do not depend on density and increase with increasing tree diameter. The survival rates for trees 1 inch and 35 inches

246

in diameter are 0.92 and 0.99, respectively. Since Douglas fir is less tolerant of shade than white and red-fir, equation (5.9) is likely to underestimate true-fir survival.

Analysis of data on tree survival in other regions of the U.S. has shown that survival is related to stand density, tree vigor (growth rate), and individual tree competition in addition to tree size (see, for example, Hamilton, 1986). Including these variables in a survival model is especially important when projecting managed stands because of the ability to manipulate stand density and to control the vigor of the standing trees.

Observations of the height growth of saplings growing in openings that were created by cutting overstory trees indicated that white fir and red fir can grow to 6 feet in less than 15 years (Gordon, 1973). Thus, we assume that saplings of both species grow into the next stage independently of stand density; that is, $p_i = 1.0$, $i = 1, 2, 3, 22$, 23, 24. In the same study, Gordon observed that about 30% of the white-fir and red-fir saplings died during a 5-year interval. Thus, we assume that $s_i = 0.7$, $i = 1, 2$, 3, 22, 23, 24. The sapling growth and survival observations were made in openings greater than 8 acres in size. In uneven-aged stands, openings are ususally less than 1 acre, and as a result competition from neighboring mature trees is likely to affect growth. Thus, the density-independent models defined here may overestimate the sapling growth and survival rates found in uneven-aged stands.

The regeneration vector ϕ in growth equation (5.7) contains the ingrowth functions $\phi_1(\mathbf{y})$ and $\phi_{22}(\mathbf{y})$ representing the numbers of trees added to the smallest white-fir and red-fir stages, respectively. Each ingrowth function has the same definition as the stock-recruitment functions for the fishery models that were described in Section 3.2; each is the product of the seedling establishment rate and the

247

total number of seeds produced:

$$\phi_i = \begin{cases} \psi_i(\mathbf{y}) \sum_{j=1}^{21} \tau_j x_j, & \text{if } i = 1; \\ \psi_i(\mathbf{y}) \sum_{j=22}^{42} \tau_j x_j, & \text{if } i = 22; \\ 0, & \text{otherwise,} \end{cases}$$

where ψ_1 and ψ_{22} are the seedling establishment rates for white fir and red fir, respectively, and τ_j is the number of seeds produced by a tree in stage j.

Insufficient empirical data were available for us to estimate parameters for a true-fir establishment model using statistical methods. Instead, we used peaking functions and chose parameters that subjectively fit the following observations. Observations of natural regeneration in California true-fir stands show that the annual survival rate of red-fir and white-fir seedlings is approximately 35% (Gordon, 1970; Gordon, 1979). Further, the best survival rates take place with moderate amounts of shade, and red fir is slightly less tolerant of shade than white fir. The seedling establishment rates for white fir and red fir depend on y_1, the cross-sectional crown area of the stand measured at ground level. Using these data, the functions we obtained are:

$$\psi_i = \begin{cases} \frac{0.0711 y_1}{43560} (1.0 - \frac{y_1}{43560})^3, & \text{if } i = 1; \\ \frac{0.1220 y_1}{43560} (1.0 - \frac{y_1}{43560})^5, & \text{if } i = 22. \end{cases}$$

The white-fir establishment rate peaks at 25% crown cover with a 5-year survival rate of 0.75, and the red-fir establishment rate peaks at 18% crown cover with a 5-year survival rate of 0.81.

We subjectively selected a model for seed production based on the observation that the number of seeds per mature tree produced during a 5-year interval varies between 20,000 and 50,000 (Gordon, 1979). We assumed that trees less than fifteen inches in diameter do not produce seed and that the number of seeds produced per tree increases up to 20,000 for 35-inch trees. Thus we estimate that

$$\tau_j = \begin{cases} 41.3(d_j - 13.0)^2, & \text{if } j = 11, \ldots, 21, 36, \ldots, 42; \\ 0, & \text{otherwise.} \end{cases}$$

Although we were limited by available data, researchers have compiled large data sets and estimated regeneration models for conifers in other regions of the U.S. (see, for example, Ferguson *et al.*, 1986).

Stand growth and yield projections may differ between the stage-class model that we construct and the single-tree simulator, CACTOS, because of two simplifying assumptions in the stage-class model. First, in the stage-class model, trees are classified into diameter and species classes, and each class represents a growth stage that is further described by average tree height, volume, and crown attributes. These class attributes are fixed for the course of the projection. Second, trees are assumed to be uniformly distributed across each diameter class, and a proportion of the trees in each stage are projected to grow into the next stage using a density-dependent diameter growth model. In contrast to the stage-class model, CACTOS includes density-dependent models for projecting the diameter, height, and crown dimensions of each tree record. These additional models and the fact that they can be applied to 500 tree records should allow CACTOS to make more accurate projections of stand diameter distributions and volumes over time.

In the next subsection we compare projections from these two kinds of simulators for a wide range of management prescriptions that were applied to hypothetical white-fir stands to see if the differences in model structure between the two approaches cause differences in model projections.

5.2.7 Comparison of Model Projections

An evaluation of the performance of STAGE and CAC-TOS is not complete without a comparison of stand projections to actual stand growth and yield observations. Unfortunately, we did not have repeated observations of stand growth, so we devised an alternative method of comparison (see Haight and Getz, 1987a). To evaluate the performance of STAGE, we compared its projections of hypothetical stands to those made with CACTOS in two management contexts: long-term projections without harvesting, and projections using several thinning regimes. They were then compared with those made by a single-tree simulator that has been tested against real stand growth observations.

The projections described here are made for pure white-fir stands. These are compared with projections obtained from CACTOS which, in its current version, does not include an ingrowth function. Thus, model projections are compared for stands that do not contain trees less than six inches in diameter and that are assumed to be clearcut at the end of the projection period.

The long-term projection capability of STAGE relative to CACTOS was assessed using a series of 60-year projections without harvesting. Sixteen hypothetical stands were constructed to represent four kinds of stand structure, each at four different densities. The stand structures were even-aged, uneven-aged, storied, and irregular. The initial basal areas ranged between 100 and 400 square feet per acre, and the initial volumes ranged between 10,000 and 100,000 board feet (MBF) per acre. For CACTOS, each stand contained approximately five hundred trees. Each tree was described by diameter, height, crown ratio, and tree factor. For STAGE, each stand was obtained by

TABLE 5.1. Summary of mean volumes and basal areas projected by CACTOS and STAGE for 16 white-fir stands at year 0 and year 60.

| Measure | Mean density | | Differences[a] | |
	CACTOS	STAGE	Mean	SE[b]
		Year 0		
Volume[c] (MBF/acre)	41.9	42.2	0.39	0.13
Basal area (ft ^{2}/acre)	209.0	210.0	0.75	0.40
		Year 60		
Volume[c] (MBF/acre)	82.7	86.4	3.73	0.93
Basal area (ft.^{2}/acre)	329.0	326.0	−3.10	0.55

[a] Difference between CACTOS and STAGE projections for each stand.

[b] Standard error of mean differences.

[c] Volume is measured in thousand board foot units (MBF) to a six-inch top diameter (outside bark) for trees six inches in diameter at breast height and larger.

computing a diameter distribution from the corresponding CACTOS stand.

We compared the initial and sixty-year projections of volume and basal area in addition to the final diameter distributions for each pair of stands. The mean of the initial stand densities for each set of stands and the mean of the differences in stand densities are given in Table 5.1. The mean of the volume differences for the STAGE stands is less than 1% of the corresponding mean volume of the CACTOS stands in year 0. The mean of the basal area differences for the STAGE stands are also less than 1% of the mean basal area of the CACTOS stands in year 0. The differences ranged between −2 and 4 square feet per acre for basal area and between −0.2 and 0.9 thousand board feet per acre for volume.

In year 60, the mean of the stand volumes projected with STAGE is greater than the mean of the stand volumes projected with CACTOS (Table 5.1). The mean difference between STAGE and CACTOS projections of stand volume (3.73 MBF per acre) is significantly different from zero at the 5% probability level. The larger volume projections by STAGE can be explained as follows. In STAGE, average tree heights and volumes for each diameter class are fixed for the length of the simulation. In CACTOS, tree heights are updated each period as a function of stand density, and as density increases, height growth decreases. As a result, the average volume per tree in CACTOS at year 60 tends to be less than the average volume per tree assigned to the corresponding diameter class in STAGE.

The mean of the differences between STAGE and CACTOS projections of stand volumes is the average error that would result if STAGE projections are repeatedly compared to CACTOS projections. The actual error in any one comparison may be different than the mean. The 95% prediction interval for a future error in a STAGE projection of stand volume is 1.69 to 5.77 MBF per acre. The interpretation of this prediction interval is that we are 95% confident that a future error in a STAGE projection compared to a CACTOS projection will be within this interval (see Reynolds, 1984). The width of the interval (4.08 MBF per acre) is 10% of the mean of the CACTOS projections of changes in stand volume, and a future error is likely to be less than 5% of the CACTOS projection of volume growth.

Long-term observations of volume yields in white-fir stands were not available for comparing the projection performances of STAGE and CACTOS. Nevertheless, the differences between STAGE and CACTOS projections can be put into perspective by comparing them to the projection errors of a single-tree simulator that has been evaluated against actual stand growth observations. The Prognosis simulator is a single-tree distance-independent model for projecting mixed conifer stands in the Rocky Mountains (Wykoff *et al.*, 1982). Prognosis model projections have been compared to observed yields in 119 stands that were measured for periods between 10 and

60 years (Wykoff, 1985). The mean of the differences between observed and projected volume yields was 4% of the mean of the observed changes in stand volume. The width of the 95% prediction interval was 120% of the mean observed change in volume yield. Thus, any one projection of volume yield is likely to differ by up to 60% of the observed projection of stand growth. If CACTOS projections of stand volume have this level of accuracy when they are compared with actual stand volume yields, then the prediction interval for the difference in STAGE projections compared to CACTOS projections would be insignificant compared to the prediction interval for the differences in CACTOS projections compared to actual growth observations. As a result, STAGE projections would have the same overall level of accuracy as CACTOS projections.

In year 60 the mean difference between STAGE and CACTOS projections of stand basal areas (−3.1 square feet per acre) is significantly different from zero at the 5% probability level (Table 5.1). The differences in STAGE basal area growth projections are due to slightly greater mortality of trees compared with CACTOS projections of mortality. The 95% prediction interval for the differences in basal area projections between STAGE and CACTOS is −4.3 to −1.9 square feet per acre. The width of the prediction interval (2.4 square feet per acre) is 2% of the mean basal area growth projected with CACTOS (120 square feet per acre). To put this prediction interval into perspective, we compare it with the 95% prediction interval for basal area projections computed for the Prognosis model. The width of this prediction interval (54 square feet per acre) was 148% of the mean observed change in basal area (Wykoff, 1985). If CACTOS projections of stand basal area have this level of accuracy when they are compared with observed stand basal areas, then the prediction interval for the difference in STAGE projections compared with CACTOS projections would be insignificant compared to the prediction interval for the differences in CACTOS projections compared with actual growth observations.

Statistical tests, such as the *chi-squared goodness of fit*, cannot be used to compare the differences between diameter distributions generated by both simulators because, for each simulator, the number of trees projected in each class is not independent of the number of trees projected to be in other classes. The graphical comparison of the projected diameter distributions after year 60 for an uneven-aged stand shows that both models predict the same trends in the movements of trees between diameter classes (Figure 5.3). The stage-class model projects a smoother diameter distribution because trees are assumed to be uniformly distributed across each diameter class. In CACTOS, each tree class is described with a point estimate of diameter, and after sixty-year projections, these estimates tend to be distributed nonuniformly across the two-inch diameter classes.

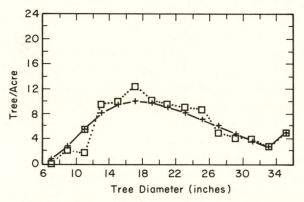

FIGURE 5.3. Diameter distributions for an uneven-aged stand after 60 years of projection with CACTOS (□) and STAGE (+).

The execution times required to make sixty-year projections varied considerably between the two simulation models. Both programs were compiled and executed using MICROSOFT FORTRAN on an IBM PC-XT microcomputer. While the stage-structured model required around ten seconds to make a sixty-year projection, CACTOS required around 360 seconds.

CACTOS and STAGE were also compared in a management setting. We defined sixteen alternative thinning regimes that varied according to thinning type, thinning intensity, and thinning interval to approximate the range of thinning regimes that would be evaluated in an optimization model. Thinning type refers to the cutting priority assigned to trees in different diameter classes. Thinning from above removes the largest trees first, and thinning from below removes the smallest trees. The sixteen regimes were divided among three types. Type 1 included regimes that removed all trees greater than a specified diameter (17, 21, or 25 inches) in each entry. Type 2 included regimes that removed a specified percentage of the stand basal area (20% or 40%) while thinning from below in each entry. Thinning type 3 included regimes that cut 30% of the stand basal area from below in the first entry, and removed trees greater than 17, 21, or 25 inches in diameter in subsequent entries. Regimes in each type had 20- and 30- year cutting cycles, and all regimes terminated in year 60 with a clearcut.

CACTOS and STAGE were used to compute yields in four different stands, which had even-aged, uneven-aged, two-aged, and irregular structures. For each stand, each of the sixteen pairs of regimes were compared by computing the differences in yield by product class and time period. We defined three product classes that depend on the top diameter of logs. Class 1 includes logs between 5 and 7 inches, class 2 contains logs between 7 and 19 inches, and class 3 contains logs greater than 19 inches. The prod-

TABLE 5.2. Projections of log and volume yields by product class and time period made with CACTOS and STAGE.

| Year and simulator | Logs per acre[a] | | | | Scribner board-feet per acre (thousands) | | | |
| | Product class[b] | | | | Product class | | | |
	1	2	3	Total	1	2	3	Total
20								
CACTOS	53	145	1	199	0.66	13.56	0.30	14.42
STAGE	50	160	3	213	0.25	15.28	0.71	16.42
40								
CACTOS	55	171	0	226	0.73	15.88	0.00	16.61
STAGE	54	171	3	228	0.28	16.54	0.81	17.66
60								
CACTOS	93	243	0	336	1.15	20.29	0.00	22.07
STAGE	88	208	4	300	0.59	17.96	0.98	19.53

[a] Harvested trees are sectioned into 16.5 foot logs starting from the stump end of the tree.

[b] Product classes 1, 2, and 3 include logs with top diameters between 5 and 7 inches, between 7 and 19 inches, and greater than 19 inches, respectively.

uct classes were defined to represent stumpage price differences associated with different-sized logs. Yields were compared using the number of logs and Scribner board-foot volume harvested.

The yields by time period for the pair of management regimes in Table 5.2 are an example of relative performance. Except where noted, the results for this pair of regimes also hold for the other stands and management regimes. The management regimes in Table 5.2 harvested all trees greater than seventeen inches in diameter on a twenty-year cycle from an even-aged stand. The differences in number of logs in product classes 1 and 2 by time period are less than 15% of the CACTOS projections. In product class 3, which contained the fewest number of logs, STAGE predicted up to four logs more than CACTOS predicted. STAGE tended to overpredict relative to CACTOS (up to fifteen logs in the largest product class).

Volume differences by product class and time period are less than 5,000 board feet. Within each time period the differences in yields across the product classes tend to cancel so that log and volume yield differences are less than 13% of the CACTOS projections. Finally, the differences in number of logs and board-foot volume produced over the entire regime are less than 5% of the CACTOS projections.

The significance of the differences in product yields for each stand depends on how stumpage price varies with product class. When price per unit volume is independent of product class, the difference in present value for each pair of management regimes is small. With a $50.00 per thousand board feet stumpage price and a 4% real discount rate, the differences in present value are less than 10% of the corresponding CACTOS projections. Within each thinning type, the rankings of the present values of the two sets of management regimes are the same. Further, both simulators produce the same rankings of the top six management regimes taken across the three thinning types. Thus, when product differentiation is not important, both simulators indicate the selection of the same optimal management regime.

When price per unit volume depends on the product class, the difference in present value for each pair of management regimes can be large. Present values for the pair of regimes in Table 5.2 were computed with product prices of $25.00, $50.00, and $100.00 per thousand board feet for product classes 1, 2, and 3, respectively, and with a 4% real discount rate. The difference in present values is 11% of the CACTOS projection. For each stand the differences in present value for the sixteen pairs of management regimes varied between 0% and 20% of the corresponding CACTOS projections. While the rankings of management regimes within each thinning type were the same, the differences in present value caused differ-

257

ent rankings of the top six regimes taken across the three thinning types. Thus, when product differentiation is important in financial analysis and a wide range of thinning regimes is considered, the STAGE simulator may give a different estimate of the optimal management regime than CACTOS.

We found for management regimes in which trees are removed from the largest diameter classes, that, relative to CACTOS, STAGE overpredicted the diameter growth and subsequent yields in the largest product class. This difference in STAGE resulted from underestimating the total crown cover, which is used in diameter growth projection. In STAGE, the crown area of trees in each diameter class is constant for the length of the simulation. In CACTOS, crown areas are updated periodically as a function of stand basal area, and crown areas expand rapidly as stand basal area decreases due to thinning from above. Thus, in these types of management regimes, the total crown area of the stand tends to be greater in CACTOS projections, which, in turn, reduces the diameter growth rates of the residual trees. It should be noted that this difference can be avoided by modifying the stage-structured model to include the impacts of changes in stand density on the average height, volume, and crown attributes of trees in each growth stage. Relaxing the assumption of constant stage-class attributes makes the stage-class model very similar in structure to the single-tree simulator; however, these refinements should not change the relative efficiency of the two simulators.

In summary, the stage-class modeling framework provides a method for simplifying a single-tree simulator with minimal loss of biological detail. There are, however, simplifying assumptions in the stage-class model that may cause differences in model projections relative to a single-tree simulator. These differences appear to be slight, however, in comparison to the demonstrated accuracy of single-tree simulators.

Single-tree simulators such as CACTOS are not suited for management analysis involving optimization because projecting the attributes of a large number of tree records requires too much execution time and storage. In contrast to the single-tree simulator, a stage-class model contains fewer state variables and growth equations, and as a result, it is computationally more efficient. This difference becomes extremely important if the models are embedded in iterative optimization routines, as are used in the management studies described below.

5.3 EVALUATING HARVEST SYSTEMS

5.3.1 Silvicultural Policy

Forest managers define silvicultural policy for existing and future stands by selecting one of two systems for long-term timber production: even-aged or uneven-aged management. Even-aged management involves the application of the clearcut, seed-tree, or shelterwood regeneration system and the establishment of an even-aged stand through planting or natural regeneration. Uneven-aged management involves cyclically harvesting trees from various size classes while creating conditions for natural regeneration using the uncut trees as sources of seed. Uneven-aged stands include trees in three or more age classes.

The adoption of one of these timber-management systems is often based on an evaluation of their relative efficiency at the stand level, in addition to the silvicultural requirements of the species involved. For example, results from efficiency analyses at the stand level have been used by the U.S. Forest Service to justify the kinds of prescriptions that are incorporated into harvest scheduling problems at the forest level. The Forest Land and Resource Management Plan for the Sierra National Forest stated that even-aged management is the desired silvicultural method for any stand in which the resource objective is the cost-effective production of timber products (USDA Forest Service, 1981, p. 199). This result, in addition to silvicultural considerations, led the Sierra National Forest to consider only even-aged management alternatives when scheduling harvests on lands where timber production was the primary objective.

Recent literature in forest management has addressed the problem of evaluating and comparing the efficiency of even-aged and uneven-aged management systems, and comparisons are made on the basis of simulated yields from steady-state management regimes (i.e., a repeated sequence of even-aged stands managed with the clearcut system and a repeated sequence of single-tree selection harvests from a sustainable, uneven-aged diameter distribution). For example, using a simulator for Wisconsin northern hardwoods, Hasse and Ek (1981) compared mean annual increments measured in various physical units for even-aged stands with the corresponding yields from steady-state uneven-aged management regimes. Chang (1981) computed a steady-state uneven-aged management regime that maximized land expectation value (LEV), and he compared the LEV associated with uneven-aged management to LEVs computed for alternative plantation regimes.

These analyses did not consider the more general problem of managing a given acre of forest land that is currently occupied by an existing stand. For even-aged management we split this problem into conversion and plantation components: determining the timing and intensity of silvicultural treatments for the current stand and determining the time when the stand is clearcut and replaced with a plantation, and then determining the timing and intensity of silvicultural treatments and clearcut age for the plantation. For uneven-aged management the problem involves determining the sequence of selection harvests that converts the current stand to steady-state uneven-aged management.

In this section we describe a stand investment model that allows the comparison of management regimes that fit these definitions of even-aged and uneven-aged management. The investment model, first described by Haight (1987), is structured to find the sequences of diameter-class harvesting rates and planting intensities that maximize the present value of an existing stand. By placing constraints on planting and harvesting, optimal even-aged and uneven-aged management regimes can be obtained, and measures of economic efficiency can be compared. Further, the investment model includes the problems of converting an existing stand to plantation management and converting the stand to steady-state uneven-aged management as special cases. Therefore, the model allows the determination of a management regime that has a present value at least as great as the present values of regimes that fit these constrained definitions of stand management.

For simplification, we limit the discussion of even-aged management to the case where existing stands are eventually replaced with plantations. We did not attempt to model the seed-tree or shelterwood regeneration methods by directly placing constraints on the general investment model as we do for conversion to plantation management. Solutions to the unconstrained investment model may resemble the seed-tree or shelterwood regeneration systems if these are the most economically efficient ways to manage the stand.

5.3.2 Investment Model

The general stand management problem is a discrete-time optimal control formulation that includes control variables for the numbers of trees harvested from diameter classes and the number of trees planted. Recalling definitions from Section 5.2, we define $\mathbf{x}(t) \in R^n$ as a state vector where $x_i(t)$ represents the number of trees in growth stage i at the beginning of time period t. We define a control vector $\mathbf{u}(t) \in R^n$ where u_i represents the number of trees harvested from the ith growth stage at the end of time period t. The elements u_i, $i = 1, \ldots, n$, belong to some suitably defined set $U \in R^n$ where

$$U = \{\mathbf{u} \in R^n \mid 0 \le u_i \le \overline{u}_i, \; i = 1, \ldots, n\}, \qquad (5.10)$$

and $\overline{u}_i$ is the number of trees in stage i immediately before harvesting at the end of period t. In addition to the harvest control, we define a planting control vector $\mathbf{v}(t) \in R^n$ where $v_i(t)$ represents the number of trees added to growth stage i at the end of period t. We constrain $v_i = 0$, $i = 2, \ldots, n$; that is, we allow planting into the smallest growth stage ($v_1 \ge 0$). More general, however, nonzero v_i would allow us to consider planting older trees, which may be considered when seedlings have a high mortality rate. We depict the planting control separately from the harvest control because there are different cost and revenue functions associated with each.

Recall the nonlinear growth dynamics from equation (5.7), which includes the growth and survival matrix $G = PS$ and the input vector ϕ. With the planting and harvest control vectors, the stand growth dynamics are

$$\mathbf{x}(1) = \mathbf{x}(0) - \mathbf{u}(0) + \mathbf{v}(0),$$
$$\mathbf{x}(t + 1) = G\big(\mathbf{y}(t)\big)\mathbf{x}(t) + \phi\big(\mathbf{y}(t)\big) - \mathbf{u}(t) + \mathbf{v}(t),$$
$$t = 1, 2, \ldots.$$

The optimization problem for harvesting and planting a stand with initial structure $\mathbf{x}(0) = \mathbf{x}_0$ is formulated to seek values of the control variables $\mathbf{u}(t)$ and $\mathbf{v}(t)$, $t = 0, 1, \ldots,$

that maximize the present value of the initial stand

$$\max_{\{\mathbf{u}(t),\, \mathbf{v}(t) | t \geq 0\}} J(\mathbf{x}_0) = \sum_{t=0}^{\infty} \delta^t \left[R\big(\mathbf{x}(t), \mathbf{u}(t)\big) - C\big(\mathbf{v}(t)\big) \right],$$

$$(5.11)$$

where R is the net value of harvested timber and C is the site preparation and planting cost. This formulation assumes that revenue and cost functions are constant over an infinite time horizon. As in Chapter 3, the discount factor δ is equal to $1/(1 + r)$ where r is a positive annual discount rate.

5.3.3 Even-aged Management

Numerical studies of even-aged stand management have either solved for the optimal management regime for an existing stand that terminates with clearcut and initiation of plantation management (e.g., Roise, 1986, Bullard *et al.*, 1985) or solved for the optimal infinite-series plantation regime (e.g., Haight, Brodie, and Dahms, 1985). These problems can be solved simultaneously by constraining the harvest and planting controls so that the infinite time horizon problem (5.11) is converted into finite time horizon conversion and plantation problems.

By definition, a plantation is established after a clearcut at the end of period K and is managed with a thinning regime during an L-period rotation. For planting density $\mathbf{v}(K)$ and harvest levels $\mathbf{u}(t)$, $t = K + 1, \ldots, K + L$, define $Q_{K,L}$ as the present value of a plantation regime that begins at the end of time period K and grows for one rotation

$$Q_{K,L} = -\delta^K C\big(\mathbf{v}(K)\big) + \sum_{t=K+1}^{K+L} \delta^t R\big(\mathbf{x}(t), \mathbf{u}(t)\big), \qquad (5.12)$$

where the growth dynamics are

$$\mathbf{x}(K + 1) = \mathbf{v}(K),$$

263

$$\mathbf{x}(t+1) = G\big(\mathbf{y}(t)\big)\mathbf{x}(t) + \phi\big(\mathbf{y}(t)\big) - \mathbf{u}(t),$$
$$t = K+1, \ldots, K+L, \qquad (5.13)$$

and the clearcut constraint at the end of the rotation is

$$\mathbf{u}(K+L) = G\big(\mathbf{y}(K+L)\big)\mathbf{x}(K+L) + \phi\big(\mathbf{y}(K+L)\big).$$

Note that the planting control is zero except for the end of period K when the plantation is established.

If we assume that an L-period plantation regime is applied repeatedly in perpetuity, the present value of the infinite series of plantations established at the end of period K is $Q_{K,L}/(1-\delta^L)$. This result is consistent with the previous assumption that prices, costs, and interest rates are constant over an infinite planning horizon.

With this definition of plantation management, the problem of converting an existing stand in K time periods to an infinite series of L-period plantations is the $K+L$ period problem,

$$\max_{\{\mathbf{u}(t),\, t=0,\ldots,K+L;\, \mathbf{v}(K)\}} J_{K,L}(\mathbf{x}_0)$$

$$= \sum_{t=0}^{K} \delta^t R(\mathbf{x}(t), \mathbf{u}(t)) + \frac{Q_{K,L}}{1-\delta^L}, \qquad (5.15)$$

subject to the growth dynamics

$$\mathbf{x}(1) = \mathbf{x}(0) - \mathbf{u}(0),$$
$$\mathbf{x}(t+1) = G(\mathbf{y}(t))\mathbf{x}(t) + \phi(t) - \mathbf{u}(t),$$
$$t = 1, \ldots, K, \qquad (5.16)$$

the clearcut constraint at the end of period K

$$\mathbf{u}(K) = G\big(\mathbf{y}(K)\big)\mathbf{x}(K) + \phi\big(\mathbf{y}(K)\big), \qquad (5.17)$$

and the growth dynamics (5.13) and clearcut constraint (5.14) for the plantation. The planting control is zero during the conversion regime, and it is positive at the end of period K as defined in the plantation regime (equations (5.12) and (5.13)). Also note that, when $K = 0$, $\mathbf{u}(0) = \mathbf{x}(0)$ and problem (5.15) reduces to the plantation management problem.

Denote $J_{K,L}^*$ as the optimal solution to the conversion and plantation problem (5.15) for a given conversion period K and rotation age L. The best conversion period length and rotation age is found by maximizing $J_{K,L}^*$ over all nonnegative K and L. This problem is simplified by noting that, by the principle of optimality (see Chapter 3), the plantation management problem can be solved separately from the conversion problem. Since constraint set (5.17) states that the stand is clearcut at the end of period K, the stand structures $\mathbf{x}(t)$, $t = K + 1, \ldots, K + L$, depend on the planting density $\mathbf{v}(K)$ and are independent of the conversion regime prior to period K. Thus, $J_{K,L}^*$ is the sum of two maximization problems:

$$\max_{\{\mathbf{u}(t),\, t=0,\ldots,K\}} \sum_{t=0}^{K} \delta^t R\big(\mathbf{x}(t), \mathbf{u}(t)\big), \tag{5.18}$$

subject to growth dynamics (5.16) and clearcut constraint (5.17); and

$$\max_{\{\mathbf{v}(K);\, \mathbf{u}(t),\, t=K+1,\ldots,K+L\}} \frac{Q_{K,L}}{1 - \delta^L}, \tag{5.19}$$

subject to the growth dynamics (5.13) and clearcut constraint (5.14). Further, the plantation management problem (5.19) can be solved for any rotation age L independently of the K-period conversion problem (5.18). This result holds for any conversion length K.

Define Q_L^* as the present value of the optimal infinite-series plantation regime that is obtained by solving problem (5.19) for rotation length L. Define Q^* as the maximum of Q_L^*, $L = 1, 2, \ldots$. The K-period conversion problem depends on Q^* as follows:

$$J_K^* = \max_{\{\mathbf{u}(t),\, t=0,\ldots,K\}} \sum_{t=0}^{K} \delta^t \, R(\mathbf{x}(t), \mathbf{u}(t)) + \frac{Q^*}{\delta^K}. \qquad (5.20)$$

Since Q^*/δ^K is constant for a given K, the optimal solution value J_K^* to problem (5.20) is independent of Q^*. If J^* is the value of the optimal conversion regime over all K (that is, J^* is the maximum of J_K^*, $K = 0, \ldots, L$), J_K^* includes the term Q^*/δ^K, and the value J^* depends on Q^*/δ^K.

In summary, the problem of determining the optimal conversion and plantation management regime is a special case of the general stand management problem (5.11), where, in this special case, constraints are imposed to require a clearcut and switch to plantation management. Further, the plantation problem is independent of the conversion problem and can be solved separately. The optimal conversion period length and conversion harvest regime depend on the value of the optimal plantation regime.

The problem of valuing land and existing trees dates back to Faustmann's (1849) work on forest land valuation. Assuming that even-aged management would be practiced indefinitely, Faustmann defined *forest value* as the sum of the stand value, which is the present value of harvests taken during the conversion period, and *land expectation value* (LEV), which is the present value of an infinite series of plantations. Thus $J_{K,L}$, the present value of conversion and plantation harvesting in expression (5.15), is equivalent to Faustmann's definition of forest value, and Q_L^*, the present value of the optimal plantation regime that solves problem (5.19), is equivalent in definition to the maximum LEV for a repeated sequence of plantations with rotation L. Faustmann noted that, since plantation management starting with bare land would eventually be practiced in perpetuity, LEV is independent and separable from the present value of the conversion regime. Thus while this result is discussed above in a more rigorous manner, it was intuitively obvious to Faustmann over 100 years ago.

It should be emphasized that the separability of the plantation problem depends on the key assumptions that prices, costs, and interest rates are constant over an infinite time horizon. If the assumptions of infinite time horizon or constant economic parameters are relaxed, optimal plantation rotations will vary over time (see Hardie *et al.*, 1984), and will depend on the period in which the conversion to plantation management takes place.

5.3.4 Uneven-Aged Management

Much of the literature on uneven-aged management has focused on the determination of optimal steady-state management policies (see Section 3.4.4 for a discussion of fixed- and equilibrium-endpoint problems). This focus on steady-state management results from the traditional forestry goal of achieving a sustainable yield of timber products. In this subsection we examine the role of the steady state in a dynamic optimization problem, we contrast different methods for imposing steady-state constraints, and we deduce the impacts of achieving steady states that are determined with different criteria. In the context of the general investment model (5.11), the planting controls are set to zero in each time period, and a steady-state constraint is imposed after a transition period with length T. Dynamic harvesting problems have been formulated and solved with either equilibrium-endpoint or fixed-endpoint constraints.

Equilibrium-endpoint problems involve the determination of transition and steady-state harvest levels with equilibrium-endpoint constraints that do not require the achievement of specific target stand structures. Recalling problem (3.84), the equilibrium-endpoint problem is

$$\max_{\{\mathbf{u}(t),\, t=0,\dots,T-1;\mathbf{u}_T\}} J_T(\mathbf{x}_0) = \sum_{t=0}^{T-1} \delta^t R\big(\mathbf{x}(t), \mathbf{u}(t)\big)$$
$$+ \frac{\delta^T}{1-\delta} R(\mathbf{x}_T, \mathbf{u}_T), \quad (5.21)$$

subject to growth dynamics (5.7) for periods $t = 0, \dots, T-1$, and the terminal state $\mathbf{x}(T) = \mathbf{x}_T$, which satisfies the steady-state constraint

$$\mathbf{x}_T = G(\mathbf{y}_T)\mathbf{x}_T + \phi(\mathbf{y}_T) - \mathbf{u}_T. \quad (5.22)$$

Note that the second term on the right-hand side of problem (5.21) corresponds to $\sum_{t=T}^{\infty} \delta^t R(\mathbf{x}_T, \mathbf{u}_T)$. If $J_T^*(\mathbf{x}_0)$ is the value corresponding to the optimal solution to the T-horizon equilibrium-endpoint problem, then $J_T^*(\mathbf{x}_0)$ approximates $J^*(\mathbf{x}_0)$, the value of the infinite time horizon problem, and converges to it as $T \to \infty$. The difference between $J_T^*(\mathbf{x}_0)$ and $J^*(\mathbf{x}_0)$ can be regarded as the cost associated with the constraint that the system must be in equilibrium for $t \geq T$. The optimal equilibrium pair $(\mathbf{x}_T, \mathbf{u}_T)$ depends on $\mathbf{x}_0$, T, and δ.

Using a linear objective function and linear growth dynamics, Michie (1985) formulated and solved a linear program equivalent to problem (5.21) with $T = 2$. For comparison with the present value of converting an existing stand to plantation management, Michie (1985) defined J_T^* as *forest value*, the sum of the present values of the transition and steady-state management regimes that constitute uneven-aged management.

Fixed-endpoint problems involve the determination of a target steady state and a transition regime that reaches the target after a finite transition period. One choice for the fixed endpoint is the extremal steady state associated with the infinite time horizon problem (5.11). When stand growth dynamics are nonlinear, optimal harvesting may converge asymptotically to an extremal steady-state pair $(\mathbf{x}_\delta, \mathbf{u}_\delta)$ that depends on the discount factor δ and satisfies the steady-state condition

$$\mathbf{x} = G(\mathbf{y})\mathbf{x} + \phi(\mathbf{y}) - \mathbf{u} \tag{5.23}$$

and a set of necessary conditions derived from Pontryagin's Maximum Principle (see equation (3.82)). If an extremal steady state exists, it can be used in the fixed-endpoint problem as follows:

$$\max_{\{\mathbf{u}(t),\, t=0,\ldots,T-1\}} I_T(\mathbf{x}_0) = \sum_{t=0}^{T-1} \delta^t R\big(\mathbf{x}(t), \mathbf{u}(t)\big)$$
$$+ \frac{\delta^T}{1-\delta} R(\mathbf{x}_\delta, \mathbf{u}_\delta), \tag{5.24}$$

subject to growth dynamics (5.7) for $t = 0, \ldots, T-1$, and the terminal state $\mathbf{x}(T) = \mathbf{x}_\delta$. If $I_T^*(\mathbf{x}_0)$ is the value corresponding to the optimal solution to the T-horizon fixed-endpoint problem, then $I_T^*(\mathbf{x}_0)$ approximates $J^*(\mathbf{x}_0)$, the value of the infinite time horizon problem, and approaches it as $T \to \infty$. The additional constraints on $\mathbf{x}(T)$ imply that $I_T^*(\mathbf{x}_0)$ is less than or equal to $J_T^*(\mathbf{x}_0)$ for any finite T. Associated with the fixed-endpoint problem is the question of reachability of the target set (that is, do any controls exist that drive the system to the specified endpoint $\mathbf{x}(T)$ in the allocated time interval T; see Section 3.4.4).

A second choice for the fixed endpoint is an *investment-efficient* steady state, which has been advocated by a generation of forest economists (Duerr and Bond, 1952; Adams and Ek, 1974; Adams, 1976; Buongiorno and Michie, 1980; Chang, 1981; Hall, 1983; Bare and Opalach, 1987). Investment-efficient steady states satisfy an economic stocking criterion that equates the marginal value

269

growth percentage of the stand to the discount rate. Let the pair (x_u, u) be a steady-state solution satisfying condition (5.23) for some choice of u. Investment-efficient steady states are determined independently of the transition regime by solving a maximization problem involving the present value of the steady-state pair (that is, the term $\frac{\delta}{1-\delta} R(x_u, u)$) and a term $R(x_u, x_u)$ representing the opportunity cost of the residual growing stock. The opportunity cost is the revenue that could be obtained by clearcutting the residual steady-state growing stock x_u (that is, $u = x_u$). The maximization problem is

$$\max_u \left[\frac{\delta}{1 - \delta} R(x_u, u) - R(x_u, x_u) \right] \qquad (5.25)$$

subject to steady-state constraint (5.23). This formulation has appealed to forest economists because if the growing stock x_u is viewed as a capital investment, then (5.25) is equivalent to maximizing LEV, as defined by the Faustmann formula (Chang, 1981; Hall, 1983). As a result, the optimal solution value for problem (5.25) is used for comparison with the LEV associated with even-aged management (Chang, 1981).

Investment-efficient steady states that solve criterion (5.25) are better understood by contrasting them with steady states found by solving the equilibrium-endpoint problem (5.21) (see Haight, 1985). Investment-efficient steady states can be found by solving problem (5.21) when $T = 1$ and $R = w'u$, where $w \in R^n$ is a vector of tree prices. In this case, problem (5.21) collapses to

$$\max_{\{u(0)\}} J_1(x_0) = w'u(0) + \frac{\delta}{1 - \delta} w'u_1 \qquad (5.26)$$

subject to

$$x_1 = x(0) - u(0),$$

and the condition for the steady-state pair $(\mathbf{x}_1, \mathbf{u}_1)$

$$\mathbf{x}_1 = G(\mathbf{y})\mathbf{x}_1 - \phi(\mathbf{y}) - \mathbf{u}_1. \tag{5.27}$$

Noting that $\mathbf{u}(0) = \mathbf{x}(0) - \mathbf{x}_1$, problem (5.26) can be rewritten as

$$\max_{\{\mathbf{x}_1\}} J_1(\mathbf{x}_0) = \mathbf{w}'\mathbf{x}(0) + \frac{\delta}{1-\delta}\mathbf{w}'\mathbf{u}_1 - \mathbf{w}'\mathbf{x}_1 \tag{5.28}$$

subject to steady-state constraint (5.27). In problem (5.28), $\mathbf{w}'\mathbf{x}(0)$ is the value of the initial growing stock, and the remaining two terms are equivalent to the criterion for investment-efficient steady states (problem (5.25)). Thus, if $(\tilde{\mathbf{x}}, \tilde{\mathbf{u}})$ is the steady-state solution to problem (5.28), and if $\mathbf{x}(0) > \tilde{\mathbf{x}}$, then $(\tilde{\mathbf{x}}, \tilde{\mathbf{u}})$ is an investment-efficient solution to problem (5.25). In this case, the value of the optimal solution to problem (5.28) is the sum of the value of the initial growing stock and the LEV, and as a result it is equivalent to the definition of forest value for even-aged management (see Chang, 1981).

It turns out that an investment-efficient steady state $(\tilde{\mathbf{x}}, \tilde{\mathbf{u}})$ satisfies conditions (3.82) for the extremal steady state only in the special case where the revenue function R is a linear combination of the harvest level and $\tilde{\mathbf{u}} > \mathbf{0}$ (see Haight, 1985, for details). Further, when the revenue function is not a linear combination of the harvest level, investment-efficient steady states cannot be found by solving the equilibrium-endpoint problem with $T = 1$, and they satisfy a different set of conditions than does the extremal steady state. Thus, for sufficiently large T, a solution to the fixed-endpoint problem (5.24) that is constrained to achieve an investment-efficient steady state has, in most cases, a lower present value relative to the present value of the solution to the fixed-endpoint problem that is constrained to achieve the extremal steady state.

271

The formulation for determining investment-efficient steady states may include different definitions of the opportunity cost of the residual growing stock. In expression (5.25) the opportunity cost is calculated as though the residual stand x_u is clearcut, and this is consistent with the literature on determining investment-efficient steady states. A different definition of opportunity cost evaluates only the merchantable trees, since it is not prudent to liquidate the unmerchantable trees (at a cost) when undertaking the alternative investment. Another definition attaches a positive value to unmerchantable trees since they have a positive value in the future as they become merchantable. Regardless of how opportunity cost is calculated, investment-efficient steady states are not the same as extremal steady states found by solving the infinite time horizon dynamic problem (5.11).

A third choice for the fixed endpoint is the maximum sustainable rent steady state. Recalling equation (3.83), the maximum sustainable rent problem is

$$\max_{\mathbf{u}} R(\mathbf{x_u}, \mathbf{u}) \qquad (5.29)$$

subject to steady-state constraint (5.23). If $(\hat{\mathbf{x}}, \hat{\mathbf{u}})$ is a maximum sustainable rent solution to problem (5.29), then, as shown in equations (3.82) and (3.83), $(\hat{\mathbf{x}}, \hat{\mathbf{u}})$ approaches the extremal steady-state solution to the equilibrium-endpoint problem (5.21) as $\delta \rightarrow 1$. Thus, for $\delta < 1$ and sufficiently large T, a solution to the fixed-endpoint problem (5.24) that is constrained to achieve a maximum sustainable rent steady state has a lower present value relative to the present value of the solution to the fixed-endpoint problem that is constrained to reach the extremal steady state.

5.3.5 Measuring Economic Efficiency

The above analysis has shown that the general stand investment model (5.11) includes even-aged and uneven-

272

aged harvesting problems as special cases. These special cases arise when constraints are imposed to require a clearcut and switch to plantation management after a specified time period or to achieve a steady-state uneven-aged management regime after one or more harvests. The present values of these even-aged and uneven-aged management regimes are called *forest values*, after the definition that Faustmann originally assigned to the present value of converting a stand to plantation management. Forest values for different management regimes can be compared to determine the most efficient timber management system.

Land expectation value is the present value of an infinite series of plantations, starting with bare ground. Plantation regimes that maximize LEV are independent of the existing stand and the harvesting regime that is used before the stand is clearcut. Land expectation value has no equivalent measure in uneven-aged management. To compare the economic efficiencies of the even-aged and uneven-aged management systems applied to a given stand, the present values of conversion and plantation regimes for even-aged management should be compared with the present values of transition and steady-state regimes for uneven-aged management.

Rideout (1985) discussed a measure called *managed forest value*, which he used to compare the efficiency of even-aged and uneven-aged management systems. As he defined it, managed forest value is the present value of any steady-state management policy. In the context of uneven-aged management, the present value of a steady state is $\frac{\delta}{1-\delta} R(\mathbf{x_u}, \mathbf{u})$, so the maximum sustainable rent solution to problem (5.29) maximizes managed forest value. In the context of even-aged management, managed forest value is the LEV or the present value of an infinite series of plantations starting with bare land. Since managed forest value ignores the present value of conversion harvesting, and since steady-state uneven-aged management regimes that maximize managed forest value are not related to extremal steady states for positive discount rates, managed forest value should not be used to compare the efficiency of even-aged and uneven-aged management systems.

The present values of solutions to the general stand management problem (5.11) have the highest present values. Solutions that maximize the present value of converting to plantation management or converting to steady-state uneven-aged management are constrained and therefore have present values that are less than or equal to the present values of solutions to the general management problem. The resemblance of management regimes obtained by solving problem (5.11) to either even-aged or uneven-aged management depends on the initial stand structure, in addition to the biological and economic models and parameters used in the analysis. Where maximum efficiency is the guiding objective, the evaluation of timber management systems should include a solution to the general stand management problem for each kind of stand that exists on the forest.

The concepts of even-aged management and steady-state uneven-aged management are keystones of forestry practice. To economists, the concepts of time discounting and efficient allocation of resources are central, and management regimes that are constrained to even-aged management or steady-state uneven-aged management are irrelevant unless they can be justified in an economic framework. There are, no doubt, good reasons for applying even-aged or steady-state uneven-aged management, but these reasons should be specified in the objective function of the general stand investment model so that their impacts on optimal management can be determined. Management regimes that meet preconceived notions of optimal harvest behavior, such as those found by solving the constrained investment model, are likely to be inefficient relative to solutions found by solving an unconstrained economic model.

5.4 NONLINEAR OPTIMIZATION

The even-aged and uneven aged stand management problems discussed in Section 5.3 are classic examples of discrete-time optimal control problems. The first formulations were solved with dynamic programming algorithms to estimate the impacts of changes in silvicultural costs and stumpage prices on rotation age and thinning intensity in even-aged stands (e.g., Brodie *et al.*, 1978, Brodie and Kao, 1979). In these studies, stand growth was projected with aggregate yield functions for stand volume, basal area, and number of trees, and no attempt was made to model the distribution of trees by size classes. Later studies coupled stage-class models and single-tree simulators with dynamic programming algorithms to determine optimal thinning regimes for even-aged stands (e.g., Riitters *et al.*, 1982; Haight, Brodie, and Dahms, 1985). While the growth models projected size distributions of trees, the control variables were limited to one or two aggregate measures of stand density. Harvests were then apportioned across the distribution of trees according to specified rules.

The problem of determining the best time sequence of size-class harvesting rates in uneven-aged stands was eventually solved using either linear programming (e.g., Michie, 1985) or nonlinear programming techniques (e.g., Adams and Ek, 1974). The linear programming method was applied with a fixed parameter stage-class model, but it could be extended to models where the movement of trees between growth stages is linearly related to stand density. Nonlinear optimization was first applied to an uneven-aged management problem in which stand growth was projected with a nonlinear stage-class model. Later studies demonstrated nonlinear optimization methods applied to problems of determining optimal size class thinning rates in even-aged stands in which growth was projected with either a stage-class model (Haight, 1987) or a single-tree simulator (Roise, 1986).

Nonlinear optimization methods are the most flexible for solving even-aged and uneven-aged management problems because they allow stand growth to be projected with equations that are linear or nonlinear functions of stand density. In this section we present gradient and penalty function methods for solving a T-horizon discrete-time optimal control problem in resource management. Another description of these methods can be found in Dreyfus and Law (1977, p. 102), and applications to forestry problems can be found in Haight, Brodie, and Adams (1985) and Haight and Getz (1987b).

Let $R(\mathbf{x}(t), \mathbf{u}(t))$ denote the revenue obtained during time period t and δ denote the discount factor. Recalling equation (3.68), the value of harvesting a resource from initial state $\mathbf{x}_0$ is

$$J_T(\mathbf{x}_0) = \sum_{t=0}^{T-1} \delta^t R(\mathbf{x}(t), \mathbf{u}(t)) + L(\mathbf{x}(T), T), \qquad (5.30)$$

where $L(\mathbf{x}(T), T)$ is the present value of the resource at time T. The optimization problem is to choose the harvest controls $\mathbf{u}(t) \in U$, $t = 0, \ldots, T-1$, that maximize $J_T(\mathbf{x}_0)$ subject to the general nonlinear equation for the resource growth dynamics

$$\mathbf{x}(t+1) = \mathbf{f}(\mathbf{x}(t), \mathbf{u}(t)) \qquad t = 0, \ldots, T-1, \qquad (5.31)$$

and the specified initial condition $\mathbf{x}(0) = \mathbf{x}_0$, where U is the set of feasible controls (see equation (5.10)).

A gradient method seeks to improve $J_T(\mathbf{x}_0)$ by successive approximations of the control-variable values. Starting with an initial guess $\mathbf{u}^0(t)$, $t = 0, \ldots, T-1$, the following procedure is used to determine a better sequence $\mathbf{u}^1(t)$,

$t = 0, \ldots, T - 1$. (Note that the superscript on the control variables represents the number of improvements made to $\mathbf{u}^0(t)$, $t = 0, \ldots, T - 1$.) The procedure described here is an extension of a solution method given by Dreyfus and Law (1977) for discrete-time optimal control problems. Second-order Newton-Raphson procedures and conjugate gradient methods may also be applied. Detailed discussions of the convergence properties and efficiencies of first- and second-order solution methods can be found in any nonlinear programming text (for example, Bazaraa and Shetty, 1979).

To improve $J_T(\mathbf{x}_0)$, we need to know how it behaves if we change $u_j^0(t)$ and keep all other control variables fixed. Define $P_t(\mathbf{x}(t), \mathbf{u}^0(t))$ as the present value of the resource from period t through period T, starting in period t with state $\mathbf{x}(t)$ and using the control $\mathbf{u}^0(t)$. The value function P_t satisfies the recurrence relation

$$P_t = \delta^t R(\mathbf{x}(t), \mathbf{u}^0(t)) + P_{t+1}(\mathbf{x}(t + 1), \mathbf{u}^0(t + 1)).$$

Note that $\mathbf{x}(t+1)$ can be replaced with the growth dynamics $\mathbf{f}(\mathbf{x}(t), \mathbf{u}^0(t))$ from equation (5.31) to obtain

$$P_t = \delta^t R(\mathbf{x}(t), \mathbf{u}^0(t)) + P_{t+1}[\mathbf{f}(\mathbf{x}(t), \mathbf{u}^0(t)), \mathbf{u}^0(t + 1)]. \quad (5.32)$$

The boundary condition is

$$P_T = L(\mathbf{x}(T), T). \quad (5.33)$$

Partial differentiation of equation (5.32) with respect to $u_j^0(t)$ tells us how $J_T(\mathbf{x}_0)$ changes for a small increase in $u_j^0(t)$:

$$\left. \frac{\partial P_t}{\partial u_j^0(t)} \right|_0 = \delta^t \left. \frac{\partial R}{\partial u_j^0(t)} \right|_0 + \sum_{k=1}^n \left. \frac{\partial P_{t+1}}{\partial x_k(t + 1)} \right|_0 \left. \frac{\partial f_k}{\partial u_j^0(t)} \right|_0, \quad (5.34)$$

where $|_0$ denotes that expressions are evaluated in terms of the state and control set $\{\mathbf{x}^0(t), \mathbf{u}^0(t)\}$. The first term

277

on the right-hand side of equation (5.34) is the change in revenue associated with the change in the control variable at the beginning of period t. The summation term on the right-hand side of (5.34) is the effect of the change in the control variable on the present value of future harvests. This is computed using the chain rule as the sum of the products of the partial derivatives with respect to the state variables in period $t + 1$ and the partial derivatives of the growth function with respect to the control variables in period t. Note that f_k, $k = 1, \ldots, n$ are the elements of the growth vector $\mathbf{f}$ defined in equation (5.31).

We can compute partial derivative (5.34) if we know $\partial P_{t+1}/\partial x_k(t + 1)$, $k = 1, \ldots, n$. Taking the partial derivative of (5.32), we find

$$\left. \frac{\partial P_t}{\partial x_j(t)} \right|_0 = \delta^t \left. \frac{\partial R}{\partial x_j(t)} \right|_0 + \sum_{k=1}^{n} \left. \frac{\partial P_{t+1}}{\partial x_k(t + 1)} \right|_0 \left. \frac{\partial f_k}{\partial x_j(t)} \right|_0 . \quad (5.35)$$

The first term on the right-hand side of equation (5.35) is the change in revenue associated with the change in the state variable at the beginning of period t. The summation term on the right-hand side of (5.35) is the effect of the change in the state variable on the present value of future harvests. This is computed using the chain rule as before.

We can compute the values of the partial derivatives with respect to the state and control variables by starting with the boundary condition and working backwards to the first time period. Partial differentiation of boundary condition (5.33) with respect to the state variable in period T yields

$$\left. \frac{\partial P_T}{\partial x_j(T)} \right|_0 = \left. \frac{\partial L}{\partial x_j(T)} \right|_0 . \quad (5.36)$$

For a given state and control set $\{\mathbf{x}^0(t), \mathbf{u}^0(t)\}$, we can compute the partial derivatives associated with the terminal state using equation (5.36). These are used in turn to compute the partial derivatives in period $t - 1$ using equations (5.35) and (5.34). Proceeding backwards from period $t - 1$ we can compute partial derivatives (5.35) and (5.34) for the remaining periods.

The partial derivatives are used to update the guessed control sequence $\mathbf{u}^0(t)$, $t = 0, \ldots, T - 1$, so that $J_T(\mathbf{x}_0)$ improves. Let $\Delta u_j^0(t)$ denote the change in control variable $u_j^0(t)$ so that

$$u_j^1(t) = u_j^0(t) + \Delta u_j^0(t), \tag{5.37}$$

and let $\Delta u_j^0(t)$ be proportional to the partial derivative $\partial P_t / \partial u_j^0(t)$ so that

$$\Delta u_j^0(t) = \rho \left. \frac{\partial P_t}{\partial u_j^0(t)} \right|_0, \tag{5.38}$$

where ρ is a positive scalar. Let the set F_t denote the indices of the control variables that are on the boundary of the feasible set U at time t and that become infeasible if equation (5.38) is applied; that is, for $j = 1, \ldots, n$,

$$F_t = \left\{ j \mid u_j^0(t) = 0 \text{ and } \left. \frac{\partial P_t}{\partial u_j^0(t)} \right|_0 < 0 \cup \right.$$

$$\left. u_j^0(t) = \bar{u}_j(t) \text{ and } \left. \frac{\partial P_t}{\partial u_j^0(t)} \right|_0 > 0 \right\},$$

where $\bar{u}_j(t)$ is the maximum number of trees that can be harvested from stage j at the end of period t. The change in $J_T(\mathbf{x}_0)$ associated with the change in the control variable set can be approximated:

$$\Delta J_T \approx \sum_{t=0}^{T-1} \sum_{j \notin F_t} \left. \frac{\partial P_t}{\partial u_j^0(t)} \right|_0 \Delta u_j^0(t)$$

$$\approx \rho \sum_{t=0}^{T-1} \sum_{j \notin F_t} \left(\left. \frac{\partial P_t}{\partial u_j^0(t)} \right|_0 \right)^2. \tag{5.39}$$

For a specified improvement ΔJ_T, we can use (5.39) to compute ρ from

$$\rho = \frac{\Delta J_T}{\displaystyle\sum_{t=0}^{T-1} \sum_{j \notin F_t} \left(\left. \frac{\partial P_t}{\partial u_j^0(t)} \right|_0 \right)^2}, \tag{5.40}$$

and thus use equations (5.38) and (5.37) to compute the new control sequence $\mathbf{u}^1(t)$, $t = 0, \ldots, T - 1$. Whenever a new value of a control variable is outside the feasible set U, its value is set equal to the nearest boundary. If the new control sequence improves J_T, it is used as the starting point for the next iteration. If the new control sequence does not improve J_T, we replace ΔJ_T by $\frac{\Delta J_T}{2}$ and recompute the new control sequence, repeating the process until J_T improves. The process terminates when we specify an increase ΔJ_T that is smaller than a predetermined limit and even this does not improve J_T. No improvement in J_T is possible when

$$\partial P_t / \partial u_j(t) = 0 \quad \text{for } 0 < u_j(t) < 1,$$
$$\partial P_t / \partial u_j(t) \leq 0 \quad \text{for } u_j(t) = 0,$$
$$\partial P_t / \partial u_j(t) \geq 0 \quad \text{for } u_j(t) = \overline{u}_j(t),$$

which are the necessary conditions for constrained optimization (see Bazaraa and Shetty, 1979). Since these conditions are satisfied by any stationary solution, we must compare the values of stationary solutions that are obtained for several different starting guesses for the control variable values to have some confidence that we have obtained the globally optimal solution.

The form of $L\big(\mathbf{x}(T), T\big)$, the value function for the terminal state in problem (5.30), depends on the kind of timber management problem being solved. Recall from Section 5.3 that even-aged timber management involves the determination of a thinning regime for an existing stand that terminates in a clearcut and the determination of a plantation management regime. In both problems the stand is clearcut at the end of a finite time horizon (see equations (5.18) and (5.19)). In the context of problem (5.30), the terminal value function is

$$L\big(\mathbf{x}(T), T\big) = \delta^T R\big(\mathbf{x}(T), \mathbf{x}(T)\big), \qquad (5.41)$$

where $R\big(\mathbf{x}(T), \mathbf{x}(T)\big)$ is the net revenue obtained from clearcutting stand with state $\mathbf{x}(T)$, that is, $\mathbf{u}(T) = \mathbf{x}(T)$. Thus, even-aged management problems of determining optimal control levels $\mathbf{u}(t)$, $t = 0, \ldots, T - 1$, can be solved with the gradient method described above subject to growth dynamics (5.31), feasibility constraints (5.10), and terminal state $\mathbf{x}(T)$ free.

Solutions to the general stand management problem (5.11), which has an infinite time horizon, can be approximated by repeatedly using the gradient method to solve a finite time horizon problem in which the value function for the terminal state is the clearcut value of the stand (that is, equation (5.41)). For sufficiently large T, the form of the terminal value function essentially has no effect on the optimal solution for the initial period control variables. Thus, the optimal controls for the first period are saved, the stand is grown for one period, and a finite time horizon problem for this new stand is solved. In cases demonstrating this method, Haight (1985) found that stable solutions were obained that were independent of random initializations of the control variables.

The uneven-aged stand management problems formulated in Section 5.3 have either fixed- or equilibrium-endpoint constraints (equations (5.24) and (5.22), respectively). These problems can be converted into equivalent unconstrained problems with free terminal points by adding a penalty to the terminal value function whenever the terminal state violates its steady-state constraint.

In the context of problem (5.30), the terminal value function for an equilibrium-endpoint problem is

$$L\big(\mathbf{x}(T), T\big) = \frac{\delta^T}{1 - \delta} R(\mathbf{x}_T, \mathbf{u}_T), \qquad (5.42)$$

where the terminal state $\mathbf{x}(T) = \mathbf{x}_T$ satisfies the equilibrium constraint

$$\mathbf{x}_T = G(\mathbf{y})\mathbf{x}_T + \phi(\mathbf{y}) - \mathbf{u}_T \qquad (5.43)$$

(see equation (5.22)). From equilibrium constraint (5.43), it follows that the steady-state harvest vector $\mathbf{u}_T = \mathbf{h}(\mathbf{x}_T)$ satisfies

$$\mathbf{h}(\mathbf{x}_T) = G(\mathbf{y})\mathbf{x}_T + \phi(\mathbf{y}) - \mathbf{x}_T \qquad (5.44)$$

and the constraint $\mathbf{h}(\mathbf{x}_T) \geq \mathbf{0}$. To ensure that these constraints are satisfied, let $\mathbf{q}(\mathbf{x}_T)$ be a penalty vector containing elements q_i, $i = 1, 2, \ldots, n$, that are scalar functions defined by

$$q_i(\mathbf{x}_T) = \begin{cases} 0, & \text{if } h_i(\mathbf{x}_T) \geq 0; \\ h_i^2(\mathbf{x}_T), & \text{if } h_i(\mathbf{x}_T) < 0. \end{cases} \qquad (5.45)$$

Thus, when the steady-state harvest level in stage i is nonnegative, the penalty function q_i is zero, and when the harvest level is negative, the penalty function equals the square of the harvest level. Note that q_i is a continuous, nonnegative function of the terminal state $\mathbf{x}_T$. The penalty function is added to the terminal value function (5.42) by selecting a suitable vector $\boldsymbol{\kappa} \in R^n$ of positive constants; that is, the augmented terminal value function becomes

$$L\big(\mathbf{x}(T), T\big) = \frac{\delta^T}{1 - \delta} R\big(\mathbf{x}_T, \mathbf{h}(\mathbf{x}_T)\big) - \boldsymbol{\kappa}'\mathbf{q}(\mathbf{x}_T). \qquad (5.46)$$

Note that the terminal state $\mathbf{x}_T$ must also satisfy

$$\mathbf{x}_T = G\big(\mathbf{y}(T - 1)\big)\mathbf{x}(T - 1) + \phi\big(\mathbf{y}(T - 1)\big) - \mathbf{u}(T - 1).$$

Thus, for a given vector κ, the equilibrium-endpoint problem of determining optimal control levels $\mathbf{u}(t)$, $t = 0, \ldots, T-1$, can be solved with a gradient method subject to growth dynamics (5.31), feasibility constraints (5.10), and terminal state $\mathbf{x}(T) = \mathbf{x}_T$ free.

Fixed-endpoint problems are solved by defining a penalty vector that penalizes the terminal value function whenever the desired terminal state is not reached. Let $(\hat{\mathbf{x}}, \hat{\mathbf{u}})$ represent the desired terminal steady state determined independently from the transition regime. Recalling the discussion in Section 5.3, $(\hat{\mathbf{x}}, \hat{\mathbf{u}})$ may be the extremal steady state, the investment-efficient steady state, or the maximum sustainable rent steady state. In the context of problem (5.30), the terminal value function for a fixed-endpoint problem is

$$L\big(\mathbf{x}(T), T\big) = \frac{\delta^T}{1-\delta} R(\hat{\mathbf{x}}, \hat{\mathbf{u}}), \qquad (5.47)$$

where the terminal state $\mathbf{x}(T) = \hat{\mathbf{x}}$. To ensure that this constraint is satisfied, let $\mathbf{q}(\mathbf{x}(T))$ be the penalty vector containing elements q_i, $i = 1, 2, \ldots, n$, that are scalar functions defined by

$$q_i = \big(x_i(T) - \hat{x}_i\big)^2. \qquad (5.48)$$

Note that whenever the terminal state $x_i(T)$ does not equal the desired state $\hat{x}_i$, the penalty function q_i is positive. The penalty function is added to the terminal value function (5.47) by selecting a suitable vector $\kappa \in R^n$ of positive constants; that is, the augmented terminal value function becomes

$$L\big(\mathbf{x}(T), T\big) = \frac{\delta^T}{1-\delta} R\Big(\mathbf{x}(T), \mathbf{h}\big(\mathbf{x}(T)\big)\Big) - \kappa' \mathbf{q}\big(\mathbf{x}(T)\big). \quad (5.49)$$

Since $\mathbf{x}_T$ is a function of the state and control in period $T-1$, for a given vector κ, the fixed-endpoint problem of

283

determining optimal control levels $\mathbf{u}(t)$, $t = 0, \ldots, T - 1$, can be solved with a gradient method subject to growth dynamics (5.31), feasibility constraints (5.10), and terminal state $\mathbf{x}(T)$ free. Since each q_i is minimized when $\mathbf{x}(T) = \hat{\mathbf{x}}$ (see equation (5.48)) by subtracting $\boldsymbol{\kappa}'\mathbf{q}(\mathbf{x}(T))$, as in (5.49), the maximum solution should yield $\mathbf{x}(T) = \hat{\mathbf{x}}$ provided $\hat{\mathbf{x}}$ is feasible (that is, the system can be driven from $\mathbf{x}_0$ to $\hat{\mathbf{x}}$ over $[0, T]$ using controls $\mathbf{u} \in U$; see Section 3.4.2).

In practice, it is possible to get arbitrarily close to the optimal solution to the original problem by computing the solution to the augmented problem for sufficiently large values for the elements κ_i of $\boldsymbol{\kappa}$. However, for very large values for κ_i, more emphasis is placed on feasibility and the gradient solution method moves rapidly towards a feasible point. Typically a solution is reached that is far from optimal, but movement away from the point is difficult because of the size of the penalty function, and as a result, premature termination of the solution algorithm takes place. This problem is avoided by solving a sequence of problems for increasing penalty function parameters. With each new value for $\boldsymbol{\kappa}$, the gradient method is employed starting with the optimal solution corresponding to the previous parameter problem. The solution to each problem is generally infeasible, but as the elements of $\boldsymbol{\kappa}$ are made large, the solutions approach a feasible solution to the original problem.

The sequential solution method is summarized as follows (see Bazaraa and Shetty, 1979, p. 341). To initialize the algorithm, choose a termination scalar $\varepsilon > 0$, initial guesses for the control variables $\mathbf{u}^0(t)$, $t = 0, 1, \ldots, T - 1$, initial vector of penalty function parameters $\boldsymbol{\kappa}^0 > 0$ (i.e., all elements of $\boldsymbol{\kappa}^0$ are positive), and a scalar $\beta > 1$. Set the counter $k = 0$.

1. Solve the augmented problem starting with $\mathbf{u}^k(t)$,

$t = 0, 1, \ldots, T - 1$ and $\boldsymbol{\kappa}^k$. Let the solution be $(\mathbf{x}^{k+1}(t), \mathbf{u}^{k+1}(t))$, $t = 0, 1, \ldots, T - 1$ and $\mathbf{x}^{k+1}(T)$.

2. If $\boldsymbol{\kappa}^{k'}\mathbf{q}(\mathbf{x}^{k+1}(T)) < \varepsilon$, stop. Otherwise, let $\boldsymbol{\kappa}^{k+1} = \beta\boldsymbol{\kappa}^k$, replace k by $k + 1$ and go to step 1.

The implementation of the gradient method is made easier by redefining the state and control variables. Define $\mathbf{x}(t) \in R^n$ and $\mathbf{u}(t) \in R^n$ as, respectively, the resource state and harvest control at the beginning of time period t. The harvest control $u_i(t)$ represents the percentage of the number of individuals in stage i that are harvested at time t. Thus, the elements u_i, $i = 1, \ldots, n$, belong to set $U \in R^n$ where

$$U = \{\mathbf{u} \in R^n \mid 0 \leq u_i \leq 1, \ i = 1, \ldots, n\}.$$

Defining the harvest controls as percentages is a convenient tactic that avoids control constraints written in terms of the system state (i.e., the control constraint set (5.10)). In the next section we apply the gradient and penalty function methods to optimal harvesting of California true-fir stands.

5.5   APPLICATION TO TRUE FIR

5.5.1   Case Background

Timber stands that include white and red fir are an important component of the timber supply in California. More than 3.5 million acres or 20% of California's commercial forest land is dominated by pure or mixed stands of white fir, *Abies concolor* (Gord. & Glend.) Lindl. (Iowiana [Gord.]), and red fir, *Abies magnifica* A. Murr. True-fir stands grow on sites between 5,000 and 7,000 feet in elevation on the west side of the Sierra Nevada mountains in central California and between 4,000 and 6,000 feet in the Siskiyou mountains in Northern California. True-fir lumber production between 1976 and 1984 varied between 0.5 and 1.0 billion board feet annually, which accounted for 15% to 20% percent of the California lumber production.

Both even-aged and uneven-aged silvicultural systems are used to harvest and grow true fir, and the preferred system depends on the management goal. Even-aged management using the clearcutting, seed-tree, or shelterwood regeneration systems is preferred in both young-growth and old-growth stands because this system is viewed as being the most economically efficient method of producing timber. Uneven-aged management is preferable to even-aged management when steady state timber harvests and conservation of visual quality, wildlife habitat, and soil properties are desired goals in addition to maximizing the present value of timber production.

In this section we present two case studies for stands that include mixtures of California white fir and red fir. The first case examines the costs associated with converting stands to steady-state uneven-aged management. The second case evaluates the assertion that even-aged management is the most efficient method of timber production. In both cases stand growth is projected with the stage-class model described in Section 5.2.

Assuming that trees can be harvested from any white-fir or red-fir growth stage, we define the revenue function

$$R\big(\mathbf{x}(t), \mathbf{u}(t)\big) = \mathbf{w}'\mathbf{u}(t),$$

where $\mathbf{w}$ is a vector of stumpage prices defined on a per tree basis. Trees less than 6 inches in diameter are assumed to be unmerchantable and cost 25 cents per tree to cut. The California State Board of Equalization, which reports average stumpage prices for use in yield tax calculations, priced true fir at $25.00 per thousand board feet (Scribner) for the first half of 1986. White-fir and red-fir trees greater than 6 inches in diameter were assigned this

TABLE 5.3. Stumpage prices ($/tree) for white fir (WF) and red fir (RF).

Diam. (in.)	Price ($/tree)		Diam. (in.)	Price ($/tree)	
	WF	RF		WF	RF
0[a]	−0.25	−0.25	15	2.51	2.44
1	−0.25	−0.25	17	3.97	3.95
3	−0.25	−0.25	19	5.91	5.99
5	−0.25	−0.25	21	8.40	8.66
7	0.09	0.09	23	11.50	12.05
9	0.32	0.29	25	15.28	16.25
11	0.75	0.70	27	19.81	21.36
13	1.46	1.39	29	25.15	27.48

[a] The 0-inch diameter class includes trees in three sapling stages: 0 to 2, 2 to 4, and 4 to 6 feet in height, respectively.

price, and the prices were assumed to be constant over the planning horizons. Note that, for simplification, we do not assign any stumpage price premium for growing large trees. The values for **w** are the products of tree volume and stumpage price and are listed in Table 5.3. The real discount rate is 4%.

5.5.2 Fixed- and Equilibrium-Endpoint Problems

In this case we solve the fixed- and equilibrium-endpoint formulations to determine the impacts of the transition period length and the criterion for the terminal steady state on the costs of converting to steady-state uneven-aged management (see also Haight and Getz, 1987b). Fixed- and equilibrium harvest constraints were imposed after transition period lengths that varied between 0 and 60 years with harvests allowed every 5 years. The problems were solved starting with the young-growth mixed species stand listed in Table 5.4.

Using the equilibrium-endpoint formulation (equation (5.21)), the present values of optimal transition and steady-state regimes improve with increasing transition period length and level off at $440.00 per acre, the present value of the 60-year transition regime (Figure 5.4). The present

TABLE 5.4. Initial diameter distribution for a young-growth white fir (WF) and red fir (RF) stand. The total values of the WF and RF growing stocks are, respectively, $88.80 and $68.00 per acre.

Diam. (in.)	Density (trees/acre)		Diam. (in.)	Density (trees/acre)	
	WF	RF		WF	RF
0[a]	180.0	149.0	15	9.7	2.4
1	115.0	145.4	17	8.8	2.0
3	51.3	47.1	19	8.1	1.9
5	32.2	23.2	21	5.0	1.8
7	23.2	13.7	23	0.0	1.7
9	17.0	6.4	25	0.0	1.7
11	13.7	4.0	27	0.0	1.8
13	11.3	2.9	29	0.0	0.8

[a] The 0-inch diameter class includes trees in three sapling stages: 0 to 2, 2 to 4, and 4 to 6 feet in height, respectively.

values of the 0- and 10-year transition regimes are $50.00 per acre (11%) and $46.00 per acre (10%) less than the present value of the 60-year regime. The costs of achieving steady states in years 20 to 50 are less than $25.00 per acre, which is 6% of the present value of the 60-year regime.

The steady states associated with the 0- and 60-year transition regimes have different species compositions (Table 5.5). The steady state obtained in year 60 includes only white fir. During transition harvesting, red fir is cut when it becomes merchantable, and the remaining unmerchantable red fir is cut in year 60, prior to the establishment of the steady state. The steady state obtained in year 0 includes both species. In this case, red fir is not completely liquidated because of the cost of cutting the unmerchantable trees, and merchantable red-fir trees are kept in the stand to satisfy the steady-state requirement. The steady state obtained in year 10 includes white fir in higher proportions, and the steady states obtained in years 20 to 50 include only white fir. As the transition period increases, the value of the steady-state yield

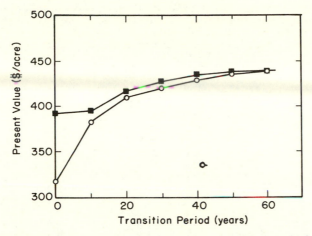

FIGURE 5.4.   Present-value vs. transition-period length for regimes that solve the equilibrium-endpoint formulation (5.21) (∎) and the fixed-endpoint formulation (5.24) (o).

decreases from \$46.90 per acre $(T = 0)$ to \$45.50 per acre $(T = 60)$.

The equilibrium-endpoint problems were relatively difficult to solve because the value function for the terminal steady state included functions for the stage-class growth dynamics (note equations (5.44) to (5.46)). The penalty function algorithm involved solving a sequence of unconstrained optimization problems with increasing penalty function parameters. We terminated the algorithm when either $\kappa' q(x_T) < .01$ or the elements κ_i of κ each exceed 10,000. In the former case, a feasible solution had been reached, while in the latter case, the algorithm had converged to an infeasible solution. The solutions obtained at termination were most sensitive to the initial guesses given to the control variables, $u^0(t)$, $t = 0, \ldots, T - 1$, and the initial size of the elements κ_i, the penalty function parameters. Random initial guesses for the control variables caused infeasible termination; however, many of the

TABLE 5.5. Steady-state diameter distributions obtained with equilibrium endpoint constraints imposed after 0 and 60 years in a mixed white-fir (WF) and red-fir (RF) stand. The cutting cycle is five years.

| Diam. (in.) | Year 0 | | | | Year 60 | |
| | WF | | RF | | WF | |
	Stock[a] (trees/acre)	Yield	Stock (trees/acre)	Yield	Stock (trees/acre)	Yield
0[b]	179.0	0.0	147.0	0.0	185.0	0.0
1	95.7	0.0	137.1	0.0	90.0	0.0
3	46.2	0.0	46.8	0.0	42.7	0.0
5	30.6	0.0	23.2	0.0	27.6	0.0
7	20.3	2.1	12.8	0.0	20.9	0.0
9	13.6	0.0	6.1	0.0	16.0	0.0
11	10.6	0.0	3.8	0.0	13.2	0.0
13	7.8	0.0	2.7	0.0	11.1	0.0
15	7.0	0.0	2.2	0.0	9.8	0.0
17	6.4	0.0	1.8	0.0	9.0	0.0
19	5.7	1.1	1.6	0.0	7.4	3.7
21	3.3	2.9	1.4	0.0	2.8	2.8
23	0.3	0.3	1.1	0.0	0.0	0.0
25	0.0	0.0	1.0	0.2	0.0	0.0
27	0.0	0.0	0.5	0.3	0.0	0.0
29	0.0	0.0	0.1	0.1	0.0	0.0
Value[c] ($/acre)	45.5	34.5	11.1	12.4	74.4	45.4

[a] Stock denotes the before-harvest stand structure.

[b] The 0-inch diameter class includes trees in three sapling stages: 0 to 2, 2 to 4, and 4 to 6 feet in height, respectively.

[c] Total undiscounted values of before-harvest growing stock and steady-state yield.

solutions were nearly feasible. Using a near-feasible solution as the initial guess for the control variables usually resulted in a feasible termination. For any near-feasible initial guess, feasible solutions were obtained with $\kappa_i^0 = 0.1$, $i = 1, \ldots, n$. Starting with penalty parameters substantially less than 0.1 resulted in infeasible termination because the penalty function did not affect the determination of the control variable values. Starting with penalty parameters substantially greater than 0.1 resulted in suboptimal solutions because the penalty function dominated the selection of the optimal control variable values.

Several near-feasible or feasible initial guesses for the control variables were explored before we were confident that an optimal solution had been obtained.

The fixed-endpoint problem given by expression (5.24) involves the determination of the optimal harvest policy for a finite transition period that terminates in a predefined steady-state policy. Three kinds of target steady states were computed (Table 5.6): the extremal steady state (ESS), the investment-efficient steady state (IESS) and the maximum sustainable rent (MSR) steady state. The criterion substantially affects the present value and species composition of transition and steady-state management.

The extremal steady state represents the harvesting regime that would be attained after a sufficient number of unconstrained harvests in any stand, assuming that biological and economic parameters are constant over time. The ESS for the bionomic model discussed above was found by solving the equilibrium-endpoint problem for the young-growth stand in Table 5.4 with transition period lengths that did not affect the resulting steady-state stand structure. The ESS includes only white-fir trees up to 21 inches in diameter, and harvests take place in the 19- and 21-inch diameter classes (Table 5.6). The value of the steady-state yield is $40.60 per acre.

The investment-efficient steady state solves the static optimization problem (5.25). This problem was solved using the penalty function method described above on the following augmented objective function that includes the steady-state harvest level $h(x_u)$ and the penalty vector $q(x_u)$ (see equations (5.44) and (5.45)):

$$\max_{x_u} \left[\frac{\delta}{1 - \delta} R(x_u, h(x_u)) - R(x_u, x_u) - \kappa q(x_u) \right]. \quad (5.50)$$

The IESS involves only red-fir trees up to 29 inches in diameter, and harvests take place in the 7- and 29-inch

TABLE 5.6. Steady states for the fixed-endpoint problem. The extremal steady state (ESS) includes white fir (WF), and the investment-efficient steady state (IESS) and the maximum sustainable rent steady state (MSR) include red fir (RF). The cutting cycle is five years.

Diam. (in.)	ESS, WF		IESS, RF		MSR, RF	
	Stock[a] (trees/acre)	Yield	Stock (trees/acre)	Yield	Stock (trees/acre)	Yield
0[b]	141.0	0.0	1032.0	0.0	714.0	0.0
1	66.3	0.0	936.6	0.0	695.0	0.0
3	34.5	0.0	320.4	0.0	222.5	0.0
5	23.8	0.0	172.1	0.0	110.6	0.0
7	18.4	0.0	38.0	15.2	66.3	0.0
9	14.4	0.0	11.9	0.0	30.9	0.0
11	12.0	0.0	8.1	0.0	19.7	0.0
13	10.2	0.0	6.2	0.0	14.5	0.0
15	9.1	0.0	5.2	0.0	11.8	0.0
17	8.5	0.0	4.6	0.0	10.2	0.0
19	6.8	5.3	4.2	0.0	9.2	0.0
21	1.1	1.1	4.0	0.0	8.7	0.0
23	0.0	0.0	3.9	0.0	8.4	0.0
25	0.0	0.0	4.0	0.0	8.4	0.0
27	0.0	0.0	4.1	0.0	8.6	0.0
29	0.0	0.0	1.8	1.8	3.4	3.4
Value[c] ($/acre)	69.8	40.6	−254.4	50.1	327.7	93.4

[a] Stock denotes the before-harvest stand structure.

[b] The 0-inch diameter class includes trees in three sapling stages: 0 to 2, 2 to 4, and 4 to 6 feet in height, respectively.

[c] Total undiscounted values of before-harvest growing stock and steady-state yield.

diameter classes (Table 5.6). The value of the steady-state yield is $50.10 per acre.

The maximum sustainable rent steady state maximizes the value of steady-state harvesting and is independent of the discount rate (see criterion (5.29)). The MSR steady state was found by solving the following augmented objective function that includes the steady-state harvest level $h(x_u)$ and the penalty vector $q(x_u)$ (see equations (5.44) and (5.45)):

$$\max_{x_u} \left[\frac{\delta}{1-\delta} R(x_u, h(x_u)) - \kappa q(x_u) \right]. \qquad (5.51)$$

FOREST MANAGEMENT

The MSR steady state involves only red-fir trees up to 29 inches, and all trees growing into the 29-inch diameter class are cut (Table 5.6). The value of the steady-state yield is $93.40 per acre.

The differences in the diameter growth model parameters for white fir and red fir (see equation (5.8)) interact with the criteria for steady-state harvesting to cause profound differences in optimal species composition and stand structure. The ESS involves only white fir because any level of stand density competition produces smaller reductions in the growth rates of white fir between 1 and 7 inches in diameter than do the same levels of competition on red-fir pole growth rates (see Figure 5.2, which plots the diameter growth competition equations for white fir and red fir). As a result, when maximum present value is the investment criterion, optimal transition harvesting involves liquidating merchantable red fir and building up the white-fir growing stock to an extremal steady state that involves only white fir.

The investment-efficient criterion seeks the steady state that maximizes the difference between the present value of an infinite series of harvests and the value of the steady-state growing stock (equation (5.25)). Since trees less than 7 inches in diameter have negative value, steady states with more unmerchantable trees have higher objective function values. Two differences in the white-fir and red-fir diameter growth equations cause the IESS to involve only red fir. First, red-fir poles less than 7 inches in diameter grow at slower rates than do white-fir poles at any level of competition (Figure 5.2). As a result, the IESS includes over 2400 unmerchantable saplings and poles that increase the value of the investment-efficient objective function. Second, red-fir sawtimber greater than 15 inches in diameter grows faster than does white-fir sawtimber for crown covers greater than 40% (see Figure 5.2). Since red-fir crown areas are smaller than white-fir crown

293

areas, more large, high-valued trees can be produced by growing red fir at high densities compared to growing white fir at the same densities.

The maximum sustainable rent criterion seeks the steady state that provides the highest value harvest and, in contrast to the investment-efficient criterion, does not assess interest costs on the steady-state growing stock. The MSR policy involves only red fir because of its superior growth rates in sawtimber diameter classes, and because red fir have smaller crown areas than white fir. As a result, more large, high-valued trees can be produced by growing red fir.

Changing the discount rate affected the structure and species composition of the extremal steady state. Discount rates between 1% and 2% resulted in steady states that included both white fir and red fir, and as the discount rate approached zero, the extremal steady state approached the maximum sustainable rent steady state which includes only red fir. Discount rates greater than 6% resulted in optimal transition regimes in which all trees were harvested before they reached seed-producing size. Optimal harvesting eventually exhausted the growing stock, and, as a result, supplemental planting would be required to maintain the stand. Exhausting the growing stock at higher discount rates was optimal in these cases because of the relatively slow growth rates of white fir and red fir on sites with moderate growth potential.

The fixed-endpoint problem (expression (5.24)) was solved for the young-growth stand in Table 5.4 using the ESS as the target stand structure after transition periods that varied between 0 and 60 years in length. Due to the more restrictive endpoint constraints, the present value of the transition and steady-state management regime is less than the present value of the equilibrium-endpoint solution for each transition period (Figure 5.4). Converting to the ESS in year 0 results in a $75.00 per acre (19%) reduction in present value relative to the equilibrium-endpoint solution because of the costs associated with harvesting unmerchantable red-fir trees in the first cut. By year 60 the management regime that converts to the ESS nearly matches the equilibrium-endpoint regime.

Feasible solutions to the fixed-endpoint problems for the young-growth stand in Table 5.4 using the IESS and the MSR steady state as targets required more than 200 years of transition harvesting. In each case, merchantable white fir was liquidated and red fir was harvested from the largest diameter class during the transition period. In each case, the present value of the fixed-endpoint regime was $264.00 per acre (60%) less than the present value of converting to the extremal steady state in 60 years.

The solutions to the fixed-endpoint problems were obtained with much less work than were solutions to the equilibrium-endpoint problems. These problems were solved with a penalty function parameter $\kappa_i^0 = 0.1$, $i = 1, \ldots, n$. Feasible termination was obtained with random starting values given to the control variables, and solutions did not depend on the initial control variable values.

5.5.3 Case Summary

When the objectives of timber harvesting include the maximization of present value and the achievement of a steady-state harvest policy, the management problem can be formulated as a dynamic harvesting model with fixed- or equilibrium-endpoint constraints. This case has focused on issues associated with these constrained problems.

For a given transition-period length, the solution to the equilibrium-endpoint problem has a higher present value than the solution to any fixed-endpoint problem, since the equilibrium-endpoint formulation places fewer constraints on the terminal steady state. The equilibrium-endpoint policy depends on the initial stand structure and

295

transition-period length, and it may differ in terms of species composition and sustainable harvest value from ESS, IESS, and MSR structures. As the transition period lengthens, the equilibrium-endpoint policy approaches the ESS, and the cost of the terminal steady-state constraint approaches zero. Numerical solutions showed that the cost of the terminal steady-state constraint can be large (greater than 10% of the present value of the unconstrained solution) for short transition periods, but the impact decreases rapidly as the transition period lengthens.

The cost of a fixed steady-state constraint depends on the criterion used to determine the target steady state and the transition-period length. With an ESS target, the solution to the fixed-endpoint problem approaches the solution to the equilibrium-endpoint problem as the transition period lengthens, and the cost of the steady-state constraint approaches zero. Numerical results showed that the cost of the ESS target in short transition periods can be large (greater than 25% of the present value of the unconstrained solution). The costs of achieving the IESS or the MSR policies can be severe (greater than 60% of the present value of the unconstrained solution) regardless of the transition-period length. As the discount rate approaches zero, the ESS policy approaches the MSR policy, and the cost of achieving the MSR policy approaches zero.

The value of the steady-state yield depends on the criterion used to determine the steady-state target. The MSR policy always provides the highest value yield, and those interested in attaining the highest value yield would want to convert to the MSR policy. In previous studies, the objective associated with the achievement of an IESS structure is the maximization of the present value of harvested yields during the transition to a steady state. It turns out that the IESS structure has the highest investment value (i.e., present value of steady-state yields net investment cost), but this criterion is not consistent with the objective of maximizing present value. In fact, this objective can be achieved more efficiently by solving the equilibrium-endpoint problem. The ESS policy may produce a relatively low sustainable yield, but this is offset by the low cost associated with achieving the ESS. The ESS has the property that, once achieved, there exists no transition policy away from the ESS that improves the present value of harvesting.

Achieving steady-state harvesting has always been a desirable goal in uneven-aged management; however, as shown above, there may exist large financial costs associated with steady-state constraints. These costs occur because optimal unconstrained harvesting may approach a steady state asymptotically or not at all. More efficient solutions could be obtained by constructing an unconstrained problem that includes the costs of yield fluctuations. For example, the uneven-aged management problem could include capital stock as a state variable with increasing marginal capital adjustment costs. Capital stock could represent liquid capital available to the landowner for investment inside or outside the forestry operation. Determining the optimal harvest policy with capital adjustment costs would directly address the need to control capital fluctuations without an arbitrarily applied steady-state constraint.

In the next case we evaluate the efficiency of even-aged and uneven-aged harvesting systems for different kinds of stands, and the optimal uneven-aged management regimes are computed without imposing steady-state constraints.

5.5.4   A Comparison of Harvest System Efficiency

The purpose of this case is to evaluate and compare the efficiency of even-aged and uneven-aged harvesting systems that are applied to existing stands. As discussed in Section 5.3, we limit even-aged management to include only the clearcutting regeneration system, and thus even-aged management involves the application of conversion and plantation management policies. Plantation management involves determining the planting density, the timing and intensity of intermediate harvests, and the clearcut age (see equation (5.19)). Conversion management involves determining the timing and intensity of intermediate harvests for an existing stand and determining the time when the stand is clearcut and replaced with a plantation (see equation (5.20)). The uneven-aged harvesting problem usually involves the determination of a harvesting regime that converts to a steady state in a finite time horizon (see previous case study), but here we do away with the steady-state constraints and determine the optimal harvesting and planting sequences that solve the general stand management problem for an existing stand (see equation (5.11)).

To determine the impact of the initial stand structure on management system efficiency, we solve even-aged and uneven-aged management problems for the young-growth stand used in the previous case (see Table 5.4) and an old-growth stand described in Table 5.7. While the young-growth stand includes trees in all growth stages, the old-growth stand has no unmerchantable trees, and the average stand diameter is about 21 inches.

The comparisons of management system efficiency are made using the stumpage prices given in Table 5.3. For simplicity, these stumpage prices are the same for both even-aged and uneven-aged management. A more realistic analysis would allow the computation of stumpage prices as a function of the residual stand structure and harvested volume in each entry. Then stumpage prices for each management system would vary.

Optimal harvest regimes are computed with fixed costs for planting ($25.00/acre) and harvesting ($20.00/acre). The fixed costs account for the labor required to pre-

TABLE 5.7. Initial diameter distribution for an old-growth white-fir (WF) and red-fir (RF) stand. The total values of the WF and RF growing stocks are, respectively, $530.0 and $381.0 per acre.

Diam. (in)	Density (trees/acre)		Diam. (in.)	Density (trees/acre)	
	WF	RF		WF	RF
0^a	0.0	0.0	15	7.0	3.0
1	0.0	0.0	17	10.0	5.0
3	0.0	0.0	19	15.0	8.0
5	0.0	0.0	21	14.0	9.0
7	2.0	1.0	23	10.0	11.0
9	2.0	3.0	25	5.0	4.0
11	3.0	5.0	27	2.0	1.0
13	5.0	3.0	29	1.0	0.0

[a] The 0-inch diameter class includes trees in three sapling stages: 0 to 2, 2 to 4, and 4 to 6 feet in height, respectively.

pare and administer the planting and harvest operations, and they are assessed at each entry in both the even-aged and uneven-aged harvesting problems. For a given sequence of entry intervals in a discrete-time problem, the fixed costs do not affect the optimal planting and harvest intensities; however, the fixed costs do affect which sequence of intervals is best. For both even-aged and uneven-aged management problems, optimal management regimes were computed with fixed entry intervals that varied between 10 and 40 years. In all cases, the fixed costs resulted in an optimal entry interval of 20 years. Optimal management regimes with entry intervals not on fixed cycles yielded slightly higher present values, but they did not change the rankings of management system efficiency presented below.

In the plantation problem, planting takes place in the first period, and in the uneven-aged management problem planting may take place in any period. In both problems we assume that planted white-fir and red-fir seedlings are added to the four- to six-foot sapling stages (e.g., x_3 and x_{24} as defined in Section 5.2). Thus, planted seedlings

299

reach the 1-inch diameter class in five years compared to fifteen years required by naturally regenerated seedlings. Planted seedlings are assumed to have the same survival rates as natural seedlings.

Solutions to the general stand management problem (equation (5.30)) were obtained with the following sequential solution algorithm. The gradient method described in Section 5.4 was applied to a 140-year planning horizon problem for a given initial stand assuming that the value function for the terminal stand state was the net revenue obtained from clearcutting (see equations (5.30) and (5.41)). We terminated the algorithm when the desired change in the objective function value (ΔJ_T in equation (5.40)) was equal to 0.01, and this did not produce a control sequence that improved J_T. Tolerances less than 0.01 did not affect the resulting control sequence. A 140-year horizon was chosen because, with a 4% discount rate, the terminal value function had no affect on the determination of the initial period controls. Further sensitivity analysis showed that the gradient method terminated with the same optimal controls for the first-period harvest for any random initialization of the control variables. Thus, the optimal controls for the first period were saved, the stand was grown for one period after applying the optimal controls, and a new 140-year planning horizon problem was solved for the subsequent stand. This procedure was repeated until a 200-year management regime had been obtained. The management regimes described below were selected after examining regimes obtained from several random initializations of the control variables. The starting conditions for the control variables had little or no effect on the optimal solutions.

Optimal management of the young-growth stand involves harvesting all merchantable red-fir trees greater than or equal to seven inches in diameter at the beginning of each twenty-year cycle, leaving progressively smaller

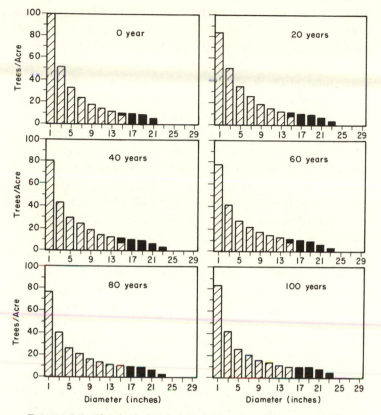

FIGURE 5.5. The shaded and unshaded regions show, respectively, the optimal number of white fir cut and the residual number of white fir by diameter class at the beginning of each 20-year cutting cycle for the young-growth stand.

numbers of red-fir saplings between one and five inches in diameter. No red fir is regenerated so that, by the end of the planning horizon, red fir has essentially been eliminated from the stand. The optimal management of the white-fir component during the first 100 years is given in Figure 5.5. At the beginning of each 20-year cycle, all white fir greater than fifteen inches in diameter are cut,

301

and trees between eleven and fifteen inches provide the seed for natural regeneration when they mature during the cycle. Natural regeneration is sufficient so that no seedlings are planted. The same harvesting behavior continues during the second 100 years (not shown). With the exception of the first cut in which a large amount of red fir is liquidated, the harvest value per period varied between \$172.00 and \$182.00 per acre. The present value of optimal uneven-aged management is \$400.69 per acre.

The optimal uneven-aged management regime for the old-growth stand involves harvesting all the red fir in the first cut, which creates a pure white-fir stand for the duration of the planning horizon. The first 100 years of white-fir management are shown in Figure 5.6. In the first four periods between years 0 and 60, optimal management involves harvesting trees from the old-growth stand when they are greater than nineteen inches in diameter. The mature trees that are left are the source of a large number of naturally regenerated seedlings and saplings that first appear in year 20 and produce a young-growth stand with merchantable trees up to fifteen inches in diameter by year 60. Harvests in years 80 and 100 take trees greater than fifteen inches while natural regeneration continues. Harvests during the second 100 years (not shown) take trees greater than fifteen inches in diameter in a manner similar to the young-growth stand management described in Figure 5.5. The harvest values vary between \$36.00 and \$836.00 per acre for the first eighty years in which the old-growth is liquidated. Subsequent harvests in the young-growth stand vary between \$165.00 and \$185.00 per acre. The present value of optimal uneven-aged management is \$920.10 per acre.

The management regime during the first 60 years in the old-growth stand is similar to the shelterwood regeneration system. In later periods, however, optimal harvesting involves uneven-aged management rather than repeating

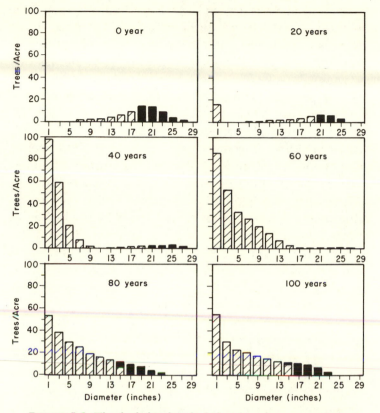

FIGURE 5.6. The shaded and unshaded regions show, respectively, the optimal number of white fir cut and the residual number of white fir by diameter class at the beginning of each 20-year cutting cycle for the old-growth stand.

the seed tree harvest cycle as classically defined. It is interesting to note that, without planting costs, planting about 500 white-fir trees per acre in the first period improves the present value of management by reducing the time required to establish the young-growth stand. A planting cost of $25.00 per acre makes this option suboptimal, however.

303

It is also interesting to note that white fir dominates uneven-aged management regimes for both young-growth and old-growth stands. In both regimes red fir is cut as soon as it is merchantable because of the joint effects of discount rate and the relative growth rates of the two species. Because of its slow growth rates in sapling and pole stages (see Figure 5.2), red fir cannot produce merchantable sawtimber yields at a rate greater than the 4% discount rate, and as a result it is more profitable to liquidate red fir and grow more vigorous white fir. At discount rates between 1% and 2%, optimal management regimes include both red fir and white fir. Discount rates less than 1% result in the liquidation of white fir with the dedication of the growing space to red fir. Red fir dominates at low discount rates because of its superior growth rates in large sawtimber stages (see Figure 5.2), and because of the small costs associated with holding large amounts of growing stock.

Even-aged management regimes were obtained by computing the best plantation regime, and then computing a sequence of conversion regimes for the young-growth and old-growth stands. Both plantation and conversion regimes were computed using the gradient method described in Section 5.4, where the terminal value function represented the clearcut value of the stand. The starting conditions for the control variables had little or no effect on the optimal management regimes.

The best plantation regime was found by solving the plantation management problem (equation (5.19)) for conversion period K equal to zero and rotation ages L that varied between 60 and 140 years in 20-year intervals. The present values of the plantation regimes peaked with a rotation of 120 years and a present value of $20.00 per acre. The optimal plantation regime involves planting 2000 white-fir trees per acre and growing them for 40 years before the first harvest (Figure 5.7). No red-fir trees are planted because of their slower growth rates during sapling and pole stages. Aproximately 50% of the planted white-fir trees die before reaching the one-inch diameter class, reflecting the rather poor survival rates of white-fir plantations in California (USDA Forest Service, 1983). Plantation thinnings take place on a 20-year cycle and trees are cut when between eleven and twenty-one inches in diameter. The trees that are clearcut in year 120 include naturally regenerated seedlings and large numbers of poles and small sawtimber that remain in the plantation.

The strategy of the plantation regime is to plant a lot of trees and harvest those that reach the larger and more valuable diameter classes first. This regime is in contrast to short-rotation plantation regimes that are often recommended for growing true fir (USDA Forest Service, 1983). We constructed several short-rotation plantation regimes and found that their present values were between 30% and 50% less than the regime given in Figure 5.7.

The present value of the plantation regime ($20.00 per acre) is low even with a relatively low regeneration cost ($25.00 per acre). Other things being equal, regeneration costs greater than $45.00 per acre would make plantation management undesirable because of its negative present value. In this case the most efficient even-aged management regime is to abandon the stand after clearcutting. The low present values found here result from the slow growth rates of seedlings and saplings. These low present values may not be repeated for true-fir stands that grow on sites with higher production capabilities.

Management regimes that convert the young-growth and old-growth stands to bare ground were computed using criterion (5.20) for conversion periods K between 0 and 80 years in 20-year intervals. The present values of the conversion regimes were added to the discounted values of plantation management to obtain the present values of conversion and plantation management (i.e., J_K^* in

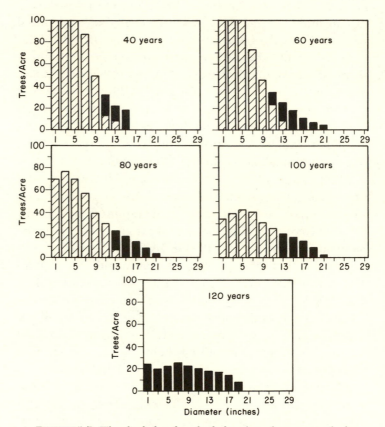

FIGURE 5.7. The shaded and unshaded regions show, respectively, the optimal number of white fir cut and the residual number of white fir by diameter class at the beginning of each 20-year cutting cycle for a white-fir plantation.

equation (5.20)). These present values are plotted against conversion-period length in Figure 5.8. Shown for comparison are the present values of the optimal uneven-aged regimes for the two stands. This comparison demonstrates the dominance of the present values of the uneven-aged management regimes that were obtained by solving the general stand management problem (equation (5.11)).

306

These results suggest that, for the economic assumptions given here, the present values of uneven-aged management in both young-growth and old-growth stands are at least as great as the present values of clearcutting these stands and starting plantations.

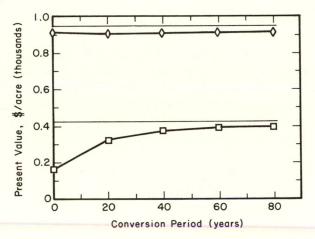

FIGURE 5.8. Present value vs. conversion period length for the young-growth (□) and the old-growth (◇) stands. The horizontal lines show the present values of uneven-aged management.

Firm conclusions about the relative profitablity of even-aged and uneven-aged management cannot be made without considering other factors that affect revenues and costs. While we used a revenue function that assumed constant marginal returns to harvesting, it is likely that the cost per unit volume harvested depends on the number and size of trees cut (that is, the more large trees are cut, the smaller the unit volume harvest costs). Such a cost function tends to concentrate harvesting in the larger tree stages, which makes uneven-aged management more profitable because clearcutting large numbers of small trees is not necessary (see Haight, 1987). On the other

307

hand, we assumed that selection harvests in uneven-aged stands do not damage the remaining trees. Increasing damage rates would tend to reduce the number of selection harvests and make even-aged management more profitable. We assumed that seedlings are established with certainty and without competition from weeds. In fact, there is much unexplained variation in seedling establishment, and weeds tend to reduce mean seedling establishment and growth rates in both uneven-aged stands and plantations. It is important to emphasize that the stage-class harvesting model can be expanded to include all of these considerations, and the stand optimization problems can be solved with the gradient methods as long as the new relations for tree growth and valuation are differentiable.

5.6   FOREST-LEVEL MANAGEMENT

The stand-level problems discussed in the previous sections addressed silvicultural questions in which harvesting depends on the evolution of trees in a narrowly defined stand. In many cases a forest property is composed of several stands that vary in age, species composition, and site quality, and harvesting decisions for any stand also depend on the management decisions made throughout the forest. In forest-level management, stand-level decisions are linked by means of constraints on the flow of harvested timber from the forest, or by economic considerations such as logging efficiency and mill capacity that apply to the forest as a whole as opposed to a single stand. In this section we describe how the stage-class model can be used to define a management problem for a forest that is composed of even-aged stands. There is a great deal of literature on this this problem (see Reed, 1986, for review). Here we only highlight some of the results and applications.

Forest dynamics can be defined using a compact stage-class model (see also Reed, 1986, and Johansson and Löfgren, 1985). Let $\mathbf{x}(t) \in R^n$ be a state vector where $x_i(t)$ represents the forest area covered by stands of age i at the beginning of period t without regard to their location. The forest model assumes that stands are even-aged. Let $\mathbf{u}(t) \in R^n$ be a vector of harvest controls where $u_i(t)$ represents the area of stands of age i that are clearcut at the end of period t, and let $\mathbf{v}(t) \in R^n$ be a vector of input controls where v_1 represents the area regenerated by planting or natural means and v_i, $i = 2, \ldots, n$ represents the area of timber in age class i that is added to the forest ownership through purchase. The areas clearcut are limited by the areas available, and the area regenerated is usually less than or equal to the total area clearcut. For an initial age-class distribution $\mathbf{x}_0 = \mathbf{x}(0)$, the forest dynamics are

$$\mathbf{x}(1) = \mathbf{x}(0) - \mathbf{u}(0) + \mathbf{v}(0),$$
$$\mathbf{x}(t+1) = S\mathbf{x}(t) - \mathbf{u}(t) + \mathbf{v}(t), \quad t = 1, 2, \ldots, \quad (5.52)$$

where S is the survival matrix (see expression (2.12). The nonzero off-diagonal elements s_i, $i = 1, \ldots, n$, of the survival matrix S are fixed proportions representing the area of stands in age class i that survive during the projection period. The survival rates are usually fixed parameters equal to one, but they also can be stochastic variables taking into account losses due to fire or other damage agents (see Reed and Errico, 1986). The forest model in (5.52) is slightly more general than the models given by Reed (1986) and Johansson and Löfgren (1985) that define input only as a function of the area clearcut. Also note that there is no density dependence in the forest model.

Analogous to the the general stand management problem (5.11), the forest-level optimization problem seeks values for the harvest and input controls that maximize the

present value of an existing forest ownership over an infinite planning horizon:

$$\max_{\{\mathbf{u}(t),\mathbf{v}(t)|t\geq 0\}} J(\mathbf{x}_0) = \sum_{t=0}^{\infty} \delta^t \left[R\big(\mathbf{x}(t),\mathbf{u}(t)\big) - C\big(\mathbf{x}(t),\mathbf{v}(t)\big) \right],$$

$$(5.53)$$

subject to the forest dynamics (5.52) and the initial forest
condition $\mathbf{x}_0 = \mathbf{x}(0)$. In the forest-level problem, R is the
net value of harvested timber and C is cost of planting and
purchasing land that contains timber.

Assuming that the revenue function is linear with respect to the harvest and input controls (i.e., there are no
positive or negative returns to scale), problem (5.53) can
be formulated as a linear program. Analysis of the linear
program shows that the optimal harvest is to clearcut each
even-aged stand when it reaches its optimal rotation age
as defined using the Faustmann formula (Johansson and
Löfgren, 1985). Thus, the management of each even-
aged stand in the linear forest does not depend on the
size and composition of the initial age-class distribution,
and optimal management regimes for each age class can
be determined independently from actions taken in stands
in other age classes. This result is the theoretical justifi-
cation for formulating and analyzing stand-level problems
without considering what takes place elsewhere in the forest.

In practice, there are many reasons why the optimal
forest-level solution cannot be found by solving independent stand-level problems. For example, harvest or cash
requirements may constrain the flow of timber volume
harvested from the forest. Soil erosion and visual-quality
standards may constrain the number of stands that can be
harvested within a watershed at one time. The revenue
function may exhibit increasing returns to scale. To han-
dle these cases the forest-level problem can often be formulated as a constrained linear program and solved with

standard linear programming techniques. The remainder of this section reviews these applications.

Harvest flow constraints were first included in the forest problem by Nautiyal and Pearse (1967) who formulated fixed- and equilibrium-endpoint problems at the forest-level that were analogous to problems (5.21) and (5.24) in uneven-aged stand management. In a linear context, these authors added fixed- and equilibrium-endpoint constraints to problem (5.53). For the fixed-endpoint problem, the target steady state was defined to have equal area covered by stands of each age class up to a maximum age defined by the Faustmann-optimal rotation. The objective was to maximize the present value of harvests that converted an irregular age-class distribution to the target forest in a specified time period. The target forest then produced a constant level of harvested timber. The authors noted that the age-class distribution of the target steady state could be made variable, and thus they formulated an equilibrium-endpoint problem and showed that its optimal solution value would be greater than or equal to the value of the solution to the fixed-endpoint problem with the same transition period. Nautiyal and Pearse also showed numerically that the present value of forest management increased monotonically with respect to the transition period, and as a result they concluded that conversion to a steady-state age-class distribution was not economically optimal.

Later applications of linear programming (e.g., the Model II formulation given by Johnson and Scheurman, 1977) converted the infinite time horizon forest problem (5.53) into a finite time horizon problem by adding a terminal value function that assigned single-stand present net values to stands left as ending inventory. Constraints that linked management actions across stands included harvest flow constraints that required period-to-period harvests to lie between certain limits or to fluctuate less than a specified absolute or proportional amount. More sophisticated versions of Model II (e.g., the FORPLAN program, which is used by the U.S. Forest Service to schedule timber harvest activities and is described by Johnson *et al.*, 1980) included constraints on harvesting in areas that are allocated to wildlife and recreation uses, constraints on ending inventory, multiple timber types and silvicultural options in addition to clearcutting. Textbook applications of linear programming to a wide range of forest planning problems can be found in Davis and Johnson (1987) and Buongiorno and Gilless (1987).

Model II formulations are essentially extensions of problem (5.53) with additional linear constraints, and, as a result, problems are solved using existing linear programming packages. However, the size and complexity of some scheduling problems formulated by the U.S. Forest Service make their solution expensive (as much as $200,000 for a single forest) and their interpretation difficult. As a result, there have been attempts to exploit the structure of the problem to produce more efficient methods that give solutions that are easier to interpret. For example, Berck and Bible (1984) show that problem (5.53) with harvest flow constraints can be more efficiently solved by applying Dantzig-Wolfe decomposition to the dual problem. Reed and Errico (1986) argue that including the state variables $\mathbf{x}(t)$, $t = 0, 1, 2, \ldots$, in addition to the harvest controls as activities in the linear program reduces both programming and execution time. While these methods have proven more efficient in cases with harvest flow constraints, they need to be tested on more complex scheduling problems that are currently included in the Model II formulation.

An advantage of formulating the forest-level problem with stage-class growth dynamics is that it explicitly allows the treatment of stochastic survival rates. Reed and Errico (1986) included the possibility of catastrophic loss from fire in the forest problem by making the stage-class transition rates s_i, $i = 1, \ldots, n$ random variables, where s_i represents the proportion of area in stands with age i that survive and grow to age class $i + 1$ during the projection interval. They show that replacing the random variables with their expected values, using the certainty equivalence principle (see Section 2.4.2), allows the maximization of expected present value using standard linear programming. Further, they show how approximately optimal feedback solutions can be obtained by sequentially solving the deterministic problem with random realizations of fire occurrence and subsequent growth dynamics.

Harvest flow constraints play a central role in the linear forest by explicitly regulating the stability of volume harvested over time. These flow constraints are included as a result of policy considerations and are criticized as being inefficient (see for example, Hyde, 1976; Waggener, 1977; and Parry *et al.*, 1983). Alternative formulations of the forest problem implicitly regulate the flow of timber harvests through the use of nonlinear revenue and cost functions and variables for investment in harvest and milling capacity. Lyon and Sedjo (1983) modeled timber price as a downward-sloping function of volume harvested and showed numerically that, as they shifted the demand function outward to reflect higher values placed by society on timber utilization, the rate of drawdown of an old-growth forest increased. They found that long-run timber harvesting converged to steady states for several downward-sloping demand functions. Allard *et al.* (1986) formulated forest-level problems that jointly determined optimal timber harvest levels and investment in harvest capacity. In these cases, the rate of timber harvesting was implicitly constrained by the costs of building up initially limited milling and harvesting capacities. Solving forest-level problems with nonlinear objective functions has required nonlinear optimization methods. These include the gradient technique outlined in Section 5.4 (Lyon and Sedjo, 1983), binary search techniques (Lyon and Sedjo, 1986), and quadratic programming (Johnson and Scheurman, 1977).

While the stage-class representation of forest growth dynamics in equation (5.52) is simple and compact, it assumes that stands are even-aged and only allows the analysis of regeneration harvesting. More complex management regimes that involve thinning and selection harvesting in both even- and uneven-aged stands are ignored. An alternative linear programming formulation, pioneered by Navon (1971) and called Model I by Johnson and Scheurman (1977), allows consideration of multiple harvest regimes for stands in each existing class of timber. In this formulation, alternative management regimes are computed for stands in each timber class and embedded as activities in the linear program. The optimization program determines the proportion of area in each timber type that is assigned to each management regime subject to area, volume flow, and other constraints discussed above. Nazareth (1980) showed how Model I could be solved with Dantzig-Wolfe decomposition. In his formulation, the subproblems were the stand-level thinning and clearcutting problems that were described for even-aged stands in Section 5.3, where the revenue functions included the shadow prices associated with forest-level constraints in the master problem. Hoganson and Rose (1984) used duality and decomposition to develop a simulation approach that approximated the optimal solution to a Model I problem. These decomposition methods could be easily extended to allow both even-aged and uneven-aged management regimes as options for existing stands. Then the relative economic efficiency of managing existing timber stands under even-aged and uneven-aged management systems subject to forest-level constraints could be determined.

5.7 REVIEW

Stand management policy is a central issue in forestry management, because stands are the basic administrative unit in which silvicultural activities such as harvesting and planting are carried out. In this chapter we described how stage-class models can be used to project the yields associated with different harvest policies and to evaluate the wide range of policies that can be formulated for multi-species and multi-aged stands.

In contrast to fisheries, the relative ease of measuring, harvesting, and planting trees has resulted in a variety of models for projecting stand growth and for managing a tree population over time. Models for predicting stand growth include simple univariate functions (Section 5.2.2), linear and nonlinear stage-class models (Sections 5.2.3 and 5.2.4), and complicated single-tree simulators (Section 5.2.5). Stand management systems vary from the use of clearcutting and the establishment of even-aged plantations to the use of selection harvesting and the maintenance of an uneven-aged stand structure.

The simplest models for stand growth are univariate functions that predict stand volume as a function of age, and these may be used to determine the clearcut age of an even-aged plantation. In practice, univariate models have found limited application because they ignore the distribution of trees by size classes. Intertree competition, even in a plantation, causes trees to have different growth rates and sizes. Stumpage price often varies by tree size so that the distribution of trees by size class affects the value of the stand. Thus, univariate models cannot be used to project the effects of size-related harvest and planting policies that control the density, growth, and value of a stand of trees.

Univariate models have been expanded to project two or more measures of aggregate stand density (such as number of trees and basal area) in a discrete-time framework. The aggregate stand attributes are then used to estimate the parameters of distribution functions (such as the Weibull) that project the distribution of trees by size classes. These models are limited because thinnings are defined by aggregate stand measures and because the forms of the distribution functions do not represent the range of possible size distributions that a stand can take. Further, the models for stand growth are applied at the aggregate stand level and do not explicitly recognize the variability in growth that can take place between trees in different size classes. These limitations become important when timber management focuses on stands that contain trees in a variety of species and age classes.

To overcome the limitations of aggregate yield models, forest modelers have constructed highly disaggregated single-tree simulators, which can project the tree attributes, such as diameter, height, and crown length, for each tree in the stand. The component models of tree growth and mortality depend on tree size as well as on various measures of stand density. Since single-tree simulators include explicit equations for tree growth and since the equations can be applied to many trees, single-tree simulators have the capability to project the widest array of stand conditions. Further, the biological detail that is included in a single-tree simulator should allow more accurate projections. However, the complexity of single-tree simulators is gained at the expense of tractability for management analysis. Single-tree simulators are not suited for iterative optimization algorithms or sensitivity analysis because projecting the attributes of a large number of trees requires too much execution time and storage, and because the number of estimated parameters in component models is too high.

The stage-class modeling framework provides a balance between biological detail and tractability for management analysis. The construction of a stage-class model in Section 5.2.6 shows that, compared with aggregate stand models, describing the stand by the frequency of individuals in different growth stages can capture the variability in stand structure that may exist in both even-aged and uneven-aged stands. Projecting stand growth using equations that predict the movement of trees between growth stages recognizes that tree growth and mortality rates vary by tree size as well as aggregate stand density.

Stage-class models, compared with single-tree simulators, provide the same degree of detail in their description of stand structure, but they do not include explicit equations for predicting the change in a wide array of tree attributes. The inclusion of growth models for an array of tree attributes undoubtably increases the subjective understanding of tree growth processes. However, the errors associated with tree-attribute models may be so great that including them does not increase the accuracy of projected aggregate stand attributes and size distributions. The comparison of projections from a stage-class model and a single-tree simulator in Section 5.2.7 suggests that the stage-class model may provide just as accurate projections with a much simpler model structure.

Stage-class models are more efficient at projecting stand structures and are analytically more tractable than single-tree simulators, because they contain fewer growth equations and state variables. Efficiency and tractability are extremely important when models are imbedded in iterative optimization routines, such as the gradient method described in Section 5.4, that are used to determine best harvest policies.

Forest managers who make long-term silvicultural policy decisions choose between even-aged and uneven-aged timber management systems. The decision is usually based in part on the relative economic efficiency of the two systems. To evaluate the efficiency of even-aged and uneven-aged management, we formulated a stand investment model

(Section 5.3) in which the objective was to find the sequence of size-class harvesting rates and planting intensities that maximized the present value of the stand over an infinite time horizon. By placing appropriate constraints on the harvest and planting controls, we obtained finite time horizon problems that fit the standard definitions of even-aged and uneven-aged management. The even-aged management problem was split into two parts: (1) determining the best thinning regime for the existing stand during the time period before it is clearcut, and (2) determining the best planting density, sequence of thinning intensities, and clearcut age of the plantation that replaces the stand after it is clearcut. Uneven-aged management involves determining the best sequence of selection harvests to undertake in the stand during a transition period that terminates with a steady-state harvest policy. Equilibrium-endpoint problems are those in which the manager specifies the time period in which a steady state is required. Fixed-endpoint problems are those in which the manager specifies the steady-state stand structure in addition to the time period in which it is required.

The formulation and solution of the general stand-investment model has allowed the evaluation and comparison of a much wider range of management regimes than have previously been considered. In the past, the efficiencies of even-aged and uneven-aged management have been estimated by subjectively extrapolating limited field experiments. The difficulties of field experimentation, however, generally preclude an exhaustive examination of a broad range of harvest policies. Thus, it is impossible to be certain that the best harvest sequence has been tested. Estimates of the efficiencies of optimal harvest regimes obtained by solving the general stand-investment model can be compared with the efficiencies of traditional harvest regimes and thus widen the array of policy options.

While the process of determining optimal management

regimes through modeling widens the range of management options, managers are correct to be cautious about implementing regimes that are based primarily on modeling results. Forestry research is just beginning to question the impacts of model uncertainty on the determination of harvest policy, and we believe that stage-class models are an appropriate framework to conduct uncertainty analyses. The construction of stage-class harvest models and the analysis of them in an optimization framework can focus managers' attention on those components of the population growth process that have the greatest effect on management-system efficiency. Sensitivity analyses can be used to determine the effects of errors in parameter estimates and model forms on optimal harvest policies. Stage-class models are superior to single-tree simulators when it comes to sensitivity analysis, because stage-class models have fewer variables, and the derivatives of the parameters, as functions of the state variables, are easier to obtain (cf. Section 3.3.3).

Finally, the stage-structured modeling framework provides a means of standardizing models for stand management analyses. The assumptions embedded in single-tree stand simulators are often not readily apparent and may differ radically from simulator to simulator. Consequently, results obtained using different stand simulators may be difficult to compare. On the other hand, if the same basic set of equations (i.e., equations (5.7)), with a relatively transparent structure, is used to model different systems, comparisons between systems are facilitated to the benefit of the forest stand management field as a whole.

CHAPTER SIX

Other Resources and Overview

6.1 HARVESTING AND CULLING LARGE MAMMALS

6.1.1 Introduction

The competing human demands for the use of terrestrial and marine wildlife habitat and for the protection and continued utilization of wildlife species have created a need for a better understanding of how large-mammal populations change over time. The population dynamics of terrestrial mammals, including deer, bear, elephant, and lion, have received attention because they are pests on domestic livestock, because they are valued as game species or tourist attractions, or because their numbers are endangered due to human encroachment. Marine-mammal populations, including seals, other pinnipeds, and whales are of interest primarily because of their economic value, competition with man for fish resources, and because exploitation has driven some marine mammal populations to near extinction.

Life-table parameters for large-mammal populations include high adult survival rates, low fecundity rates, and relatively long time periods before reproductive maturity. These factors interact to make changes in population size and age structure directly observable over time periods that are often longer than scientists or policy makers can afford to wait. An alternative way to understand large-mammal population dynamics is to incorporate these life-table parameters into an age- or stage-class modeling framework in order to predict how populations change over time in response to harvesting and encroachment. These predictions in turn add to the information pool available to wildlife managers and decision makers.

The scope of age- and stage-class models for large mammals is limited to single species population descriptions. Both analytical results and case studies have contributed to some generalizations about how life table parameters and environmental perturbations affect population dynamics. For example, life table parameters for large mammals often imply low rates of population increase and decrease, stability in age structure with moderate fluctuations in effective fecundity, low frequency oscillations in population size during the approach toward equilibrium (after environmental perturbations, such as disease or drought, have displaced the population away from the equilibrium), and greater sensitivity of population growth rates to changes in adult mortality than to other changes in life-table components. These dynamics in turn have consequences for management. For example, they improve the prospect of a population weathering unfavorable events, they delay the attainment of a steady-state after a major intervention, and they cause low resilience to adult harvesting. These features and their implications for management are demonstrated in the following two cases dealing with African elephants and grey seals. These cases by no means exhaust the examples of mammal population models in the literature (see, for example, Fowler and Smith, 1981), but they are presented here to highlight applications of stage-class models to managing wildlife populations.

6.1.2 Environmental Variability and Elephant Dynamics

An increasing human population and its demand for agricultural land have reduced the area available to all African wildlife populations, especially the wide-ranging elephant. Additional pressure from illegal ivory hunters has concentrated elephants in smaller ranges of relative safety within game parks. Because of high elephant concentrations in relatively small areas, managers feared that the elephants' increasing demand for forage would result in a progression towards habitat deterioration from which neither the elephant nor plant populations would recover. Managers were faced with the problem of determining an elephant population density that was compatible with normal environmental conditions in order to achieve population stability.

To determine a stable population density, Fowler and Smith (1973) formulated a nonlinear age-class model for the Kabalega (formally Murchison Falls National Park) population of elephants in Uganda. This population was suffering from overcrowding and habitat degradation due to restrictions in its range. The impacts of environmental degradation were modeled using equations for calf survival and adult fecundity that depend on the density of elephants as measured by number per unit area. Specifically, Fowler and Smith's (1973) model for the elephant dynamics had the form (cf. equations (3.26) and (3.48))

$$
\mathbf{x}(t+1) = \begin{pmatrix} 0 & \cdots & 0 & 0 \\ s_1 & \cdots & 0 & 0 \\ \vdots & \ddots & \vdots & \vdots \\ 0 & \cdots & 0 & 0 \\ 0 & \cdots & s_{64} & 0 \end{pmatrix} \mathbf{x}(t) + \begin{pmatrix} \phi(\mathbf{y}) \\ 0 \\ \vdots \\ 0 \\ 0 \end{pmatrix},
$$

where the elements x_i, $i = 0, 1, \ldots, 65$, represent the number of female elephants per square mile in age class i. The calf survival function $\phi(\mathbf{y})$ depends on the two-dimensional aggregation vector $\mathbf{y} = (y_1, y_2)'$, where the aggregated breeding-population variable y_1 and total-density variable y_2 are defined to be

$$
y_1(t) = \sum_{i=1}^{65} b_i x_i(t), \quad \text{and} \quad y_2(t) = \sum_{i=1}^{65} x_i(t).
$$

For the Kabalega elephant population, Fowler and Smith estimated that the survival parameters ranged in value from 0.80 (the value of parameters $s_1, s_2, s_{55}, \ldots, s_{64}$) to 0.98 (the value of parameters $s_{14}, \ldots, s_{35}$).

Calf survival, however, decreases with increasing population density (y_2): at high population densities the vegetation decreases, producing less shade, less food, poorer health of nursing females, and consequently lower calf survival rates. The specific form that Fowler and Smith chose for the calf survival function was

$$\phi(\mathbf{y}) = y_1(1 - 0.03y_2),$$

which, of course, only applies to population densities below $y_2 = 33.3$ individuals per square mile (typical densities ranged from one to ten individuals per square mile).

Adult fecundity rates b_i also depend on population density y_2: the calving interval and minimum age of reproduction both increase with increasing population density. The fecundity parameters b_i were also made to depend on the density variable y_2. Fowler and Smith (1973) estimated that at densities around one individual per square mile, the age at maturity is 11 years, while this age increases to a value of 20 years for densities around ten individuals per square mile. The actual individual calving rate b_i was assumed to have the same value for all i beyond the age of maturity, but this value varied with density y_2 from $b_i(1) = 0.25$ to $b_i(10) = 0.12$ (see Fowler and Smith 1973 for more details).

To verify this age-class model, Fowler and Smith projected the age distribution of the Kabalega population between 1945 and 1965 and compared the projected 1965 age distribution to the observed 1965 distribution. The observed 1945 population density was two elephants per square mile, and the age distribution approximated a stable population age structure. During the period between 1945 and 1965, due to human encroachment, the elephants were gradually restricted to an area less than one third the size of the original range. In the population projection this encroachment was modeled by forcing the elephant density to increase linearly from two per square mile in 1945 to ten per square mile in 1965. The 1964 age distribution projected from the observed distribution in 1945 closely resembles the observed 1965 age distribution (Figure 6.1).

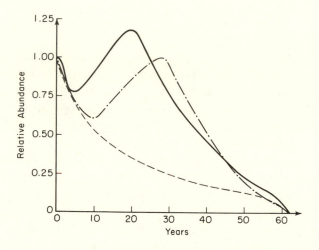

FIGURE 6.1. Observed 1965 age distribution (———) for African elephants of Kabalega National Park in Uganda. This can be compared with the age distribution predicted by Fowler and Smith (1973) for 1964 (— · —) using the observed 1945 age structure (— — —) as an initial condition for the model described in the text.

Fowler and Smith projected the 1965 age distribution forward to year 2000 assuming no further changes in the area available to the elephant population. The population density decreased from ten per square mile to two per square mile, and the age distribution appeared to be approaching a steady state. The steady-state population density computed for the density-dependent elephant population model was 1.5 per square mile. Thus, Fowler and

Smith concluded that the optimal density is less than two per square mile, which is similar to the density recommendations made by managers that are based on qualitative evaluations of elephant abundance and habitat quality. They also suggested that populations above this density should be reduced through culling because the long-lived nature of elephants results in a relatively slow approach to population steady state.

The above study modeled the effects of increasing population density on life-table parameters and associated population growth assuming continuous reductions in the area allocated to the elephant population. Working with data from the Tsavo National Park elephant population in Kenya, Croze *et al.* (1981) observed that area and consequent density changes were not the most important regulating mechanisms for population growth. Instead, they noted that forage and habitat quality changed drastically over time according to rainfall patterns. They conjectured that there exists a positive relationship between forage quality and quantity and elephant birth and survival rates. This relationship in turn can regulate population growth. Further, forage species in the semi-arid elephant habitat have evolved the capacity to recover from stressful events such as drought or excessive population density. Thus, elephant populations and their forage have self-regulatory mechanisms to prevent total destruction of their system after excessive stress.

Croze *et al.* constructed a linear model with 60 age classes for the female African elephant population on the Tsavo reserve. Thus their model was a 60×60 projection matrix L of the form given in expression (2.12), except that the survival rates of juveniles and the fecundity of young adults were assumed to vary in response to environmental conditions. Estimates of the age-specific survival rates s_i, $i = 1, \ldots, 60$, and fecundity rates b_i, $i = 1, \ldots, 60$, were obtained from observations of elephant populations in the Tsavo reserve in Kenya and Kruger National Park in South Africa. The survival rates are all close to 0.9 and the fecundity rates vary between 0.0 and 0.3. Elephants less than 6 years old are not sexually mature. Using these constant parameters, Croze *et al.* determined a stable age distribution.

Observations of the survival rates for juveniles between ages 1 and 4 years and fecundity rates for young adults between ages 8 and 16 years vary over time with the condition of the forage habitat. Sinusoidal survival and fecundity cycles with a period of 20 years were generated to model possible realizations of these parameters as they depend on the environment. These survival and fecundity rates cycled respectively between 80% and 95% of one- to four-year-olds surviving each year and eight- to sixteen-year-old females producing on average between 0.00 and 0.15 progeny each year. Croze *et al.* (1981) did not incorporate density dependence in this model. They developed the model to determine the effects of time-varying parameters on population growth.

Using the stable age structure of the Tsavo population as a starting point, 25-year projections were made with and without time-varying survival and fecundity. The time-dependent parameters did not affect the terminal age distribution, but they were related to changes in population growth rate, which were calculated as the proportional change in population size from one year to the next (Figure 6.2A,B). Thus it appears that given an age structure nearly equivalent to a stable age structure, changes from positive to negative growth in population size can be related to short-term changes in survival of the very young and conception in the younger age classes.

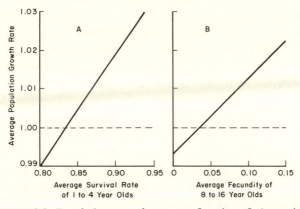

FIGURE 6.2. Population growth rate as a function of: A, survival of 1- to 4-year-olds; and B, fecundity of 8- to 16-year-old females. (Adapted from Croze *et al.* 1981.)

It is worth noting that, although the model exhibited a relation between changes in population growth rate and life-table parameters, the rate of change in population size was small. The response of the population growth rate to environment-dependent changes in life-table parameters begins the process of reducing or increasing the overall population size. Whether reductions in population size will be fast enough in cases where forage quality and quantity are diminished due to drought is not known. However, after severe reductions in plant abundance due to drought, it appears that the plant community can change its species composition and improve its abundance. Such changes in the abundances of different plant species have been observed following the severe droughts in the early 1970s in the Tsavo reserve. These modeling results and observations support the conjecture that internal mechanisms for regulating the size of both elephant and plant populations exist, and that these mechanisms

327

can help avoid the gradual depletion of the foraging habitat.

Wu and Botkin (1980) expanded the density-independent age-class model discussed above to explicitly include female reproductive stages. This is important because, while elephants breed seasonally, their gestation interval is just under two years (22 months) and, as a result, sexually mature females in the same age class may be in one of three reproductive classes: susceptible to pregnancy, pregnant less than one year, and pregnant more than one year. The reproductive life history of a mature female elephant is diagrammed in Figure 6.3, assuming the parameters pertain to a female of age i of which a proportion α_{i1} of previously susceptible females become pregnant and a proportion α_{i3} of females that have just given birth again become pregnant in the next time cycle (obviously $\alpha_{i2} = 0$ unless abortions during the first year of pregnancy are explicitly accounted for). The survival parameters s_{i1}, s_{i2}, s_{i3} refer to the three reproductive classes.

A female begins life as a calf and either dies or becomes a sexually immature juvenile. A juvenile becomes a mature adult susceptible to pregnancy, then a female pregnant for less than one year, then pregnant for more than one year, subsequently returning to the status of a susceptible female.

The female elephant life history is incorporated into the age-class model as a special case of a general approach that defines reproductive classes for each age class. In this general approach, the state vector $\mathbf{x}(t)$ is a lexicographical list of elements $x_{ij}(t)$ representing the number of individuals in age class i in reproduction class j at the beginning of time period t. Let m_i represent the number of reproductive stages for age class i. The immature age classes contain one reproductive stage (not susceptible) and the mature age classes contain the three reproductive stages

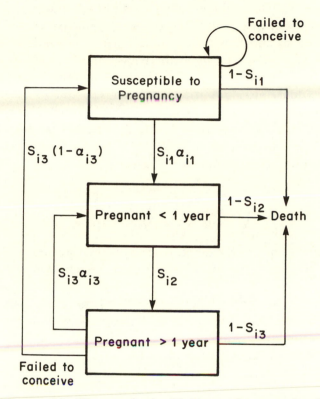

FIGURE 6.3. The reproductive life history of a mature female African elephant.

described above. The transition matrix L is now generalized to contain a top row of vector elements $\mathbf{b}_i$ and off-diagonal matrix elements S_i

$$
L = \begin{pmatrix}
\mathbf{b}_1 & \mathbf{b}_2 & \cdots & \mathbf{b}_{n-1} & \mathbf{b}_n \\
S_1 & 0 & \cdots & 0 & 0 \\
0 & S_2 & \cdots & 0 & 0 \\
\vdots & \vdots & \ddots & \vdots & \vdots \\
0 & 0 & \cdots & 0 & 0 \\
0 & 0 & \cdots & S_{n-1} & 0
\end{pmatrix},
\qquad (6.1)
$$

329

where $\mathbf{b}_i$ is an m_i-dimensional vector of fecundity parameters for elephants in the various reproductive classes of age class i and S_i is an $m_i \times m_{i+1}$ matrix of survival and transition rates describing the transition of i-year-old individuals in various reproductive states to $i + 1$-year-old individuals into appropriately related reproductive stages.

For the immature life stages of the African elephant we simply have $m_i = 1$, $\mathbf{b}_i = 0$, and $S_i = s_i$. For the mature life stages we have $m_i = 3$, $\mathbf{b}'_i = (0, 0, 0.5)$ (assuming equal sex ratios at birth) and, from Figure 6.3,

$$S_i = \begin{pmatrix} s_{i1}(1 - \alpha_{i1}) & 0 & s_{i3}(1 - \alpha_{i3}) \\ s_{i1}\alpha_{i1} & 0 & s_{i3}\alpha_{i3} \\ 0 & s_{i2} & 0 \end{pmatrix}, \qquad (6.2)$$

where, as previously mentioned, s_{ij} is the class-dependent survival rate, and α_{ij} is the class-dependent conception rate.

While the matrix L defined by (6.1) (where, for mature females, $\mathbf{b}_i$ and S_i are as defined above) is more complex than the usual Leslie matrix, the basic model still has the linear form (see equation (2.11))

$$\mathbf{x}(t + 1) = L\mathbf{x}(t),$$

which makes it amenable to analysis. For example, Wu and Botkin (1980) assumed that the parameters represent probabilities rather than proportions (see Section 2.4.2) and derived a method for calculating the relationship between the change in population size and age-dependent parameters for fecundity and mortality, a method for calculating the average number of female offspring produced over the lifetime of one female, the probability of extinction of a population, and the stable age structure.

Botkin *et al.* (1981) developed survival and fecundity parameters for the stage-class matrix (6.1) based on observations of the Wankie elephant population in Zimbabwe.

To simulate population growth and age structure, they assumed that the probabilities of conception and infant survival were linearly related to rainfall. Annual rainfall was modeled as a normally distributed random variable without serial correlation. Simulations showed that linking precipitation to infant mortality dampened population growth and kept population levels between 400 and 500 individuals over a 200-year simulation period. By contrast, linking precipitation to adult conception did not suppress growth and the population increased three- to fourfold over a 200-year simulation period. This makes sense because gestation lasts almost two years, and when a female loses a calf, she has lost two opportunities for conception. These results demonstrate the importance of including reproductive stages in the elephant population model and the importance of knowing which life-history parameters are most strongly linked to habitat quality.

The three cases discussed above demonstrate different ways to model the relationship between elephant abundance and habitat quality. In the first case this linkage was achieved using density-dependent parameters for age-class survival and fecundity. In the second two cases the linkage was made using parameters that varied over time in response to rainfall and habitat quality. An extension to these models is to include age-specific control variables for culling individuals from the population in order to avoid destruction of habitat during severe droughts or reductions in area. Another extension is to include vegetation quantity as a state variable in addition to the age distribution of elephants. Changes in vegetation state can be dependent on population density, and life-table parameters for elephants can be made directly dependent on the vegetation state.

331

6.1.3 Stability and Resilience of Harvested
Grey Seal Populations

The British population of grey seals breeds colonially on small islands within 70 km of the nearest mainland. The annual pup production of all British stocks have been estimated since the mid 1960s using direct observation on accessible islands and aerial photography on remote islands. These estimates show that the undisturbed stock increased by 6% to 7% annually between 1965 and 1975 (Harwood, 1981). At the same time it has been estimated that the seal population annually consumes more than 100,000 tons of commercially exploitable fish (Harwood, 1978). Thus, there are significant economic reasons for introducing a management policy that reduces the abundance of grey seal stock. However, because more than two thirds of the total world population of grey seal live within British waters, it would be ecologically inadvisable to reduce the population size to a point where it would become vulnerable to small perturbations in stock size. Harwood (1978, 1981) developed a density-dependent age-class model for projecting seal populations under different harvesting strategies. His goal was to determine a management policy that resulted in an equilibrium age structure that would both maintain the stock at a level considerably below its present size and be resilient to one-time changes in population size.

Based on observations of Farne Islands female grey seal population, Harwood (1978) estimated that adults 6 years or older had the same fecundity and survival parameters, so he limited his model to six age classes with the first females sexually mature at age 4. Fecundity and survival parameters were estimated from a population that was increasing at an annual rate of 7%, but it is unlikely that such a growth rate can continue indefinitely. The only density-dependent relationship that has been documented involves the survival of pups from birth to weaning at age 1. The decreasing pup survival rate is caused by a disproportionate number of deaths due to desertions and injury when the pup population is large. Harwood (1978) modeled the number of surviving pups as a Beverton and Holt recruitment curve. Thus the age class model described in Harwood (1978) is a Leslie matrix model with six age classes and a density-dependent pup survival function that has a Beverton and Holt form; that is, equations (2.5), (3.9), and (3.26) which, for the specific parameter values estimated by Harwood, can be written as

$$x_0(t) = 0.08x_4(t) + 0.28x_5(t) + 0.42x_6(t),$$

$$\mathbf{x}(t+1) = \begin{pmatrix} 0 & 0 & 0 & 0 & 0 & 0 \\ 0.93 & 0 & 0 & 0 & 0 & 0 \\ 0 & 0.93 & 0 & 0 & 0 & 0 \\ 0 & 0 & 0.935 & 0 & 0 & 0 \\ 0 & 0 & 0 & 0.935 & 0 & 0 \\ 0 & 0 & 0 & 0 & 0.935 & 0.935 \end{pmatrix} \mathbf{x}(t)$$

$$+ \begin{pmatrix} \frac{0.825x_0}{1+0.0002x_0} \\ 0 \\ \vdots \\ 0 \end{pmatrix}, \qquad (6.3)$$

where the constants determining the number of newborn pups (x_0) are the age-specific fecundity rates b_i, and the matrix elements are the age-specific survival rates s_i.

Harwood found that the equilibrium population associated with this model (see section 3.2.2) takes place when the overall pup survival rate is 0.25. Recent pup survival rates for the Farne Islands population have varied between 0.88 and 0.66. These high survival rates and the annual population growth rate indicate that the population is below its unexploited equilibrium level. Human intervention may be necessary to limit population levels and thus control their impact on the economic fisheries in the region. Care must be taken, however, to define the equilibrium population level that is not in danger of collapsing, given the underlying stochasticity in the system.

Maintaining an equilibrium through pup harvesting is attractive because pups are more easily killed than adults and a market exists for their skins. Harwood (1978) introduced two kinds of pup harvesting strategies into an equilibrium representation of the age-class growth: a constant pup quota and a fixed proportion of the pup population. Define $\mathbf{u}$ as the harvest vector where u_1 is pup quota and $u_i = 0$, $i = 2, \ldots, n$. The equilibrium population model is (this is a special case of equation (3.49))

$$\mathbf{x} = (I - S)^{-1} \begin{pmatrix} s_0 x_0 \psi(x_0) - u_1 \\ 0 \\ \vdots \\ 0 \end{pmatrix},$$

whence for given u_1 (and parameter values indicated in (6.3)) the corresponding equilibrium vector $\mathbf{x}(u_1)$ can be calculated. (Note that this equilibrium is biologically meaningful only if u_1 is sufficiently small.) The equilibrium representation of a proportional harvest is obtained by defining the matrix H to contain diagonal elements h_i, $i = 1, \ldots, 6$, representing the proportion of individuals harvested from age class i and $0 \leq h_i \leq 1$. The pup harvest proportion is h_1 and $h_i = 0$, $i = 2, \ldots, 6$. The equilibrium age-class model with a proportional pup harvest is just equation (3.35) with G replaced by S; that is,

$$\mathbf{x} = \left(I - (I - H)S\right)^{-1} \begin{pmatrix} s_0 x_0 \psi(x_0) \\ 0 \\ \vdots \\ 0 \end{pmatrix}, \qquad (6.4)$$

which, as before, can be used to find the corresponding equilibrium population level $\mathbf{x}(h_1)$.

Harwood (1981) established that the maximum sustainable yield (MSY) was the same under both harvest definitions; as expected, however, the behavior of non-MSY equilibria differed depending on how the pup harvest is defined. If the pup quota corresponds to numbers taken at MSY, then it is impossible to maintain the population at an equilibrium that has a stock level below the MSY stock level. The instability of MSY under constant harvesting is well known (for example, see Clark, 1976, 1985) and reflects the fact that stock levels below MSY do not produce enough pups to replenish a stock exploited at MSY levels. With a proportional harvest (equation (6.4)), asymptotically stable equilibria can be found that have stock levels well below the MSY level. Even though reduced equilibrium stock levels can be achieved with proportional harvests, Harwood (1978) pointed out two drawbacks. At reduced stock levels, the number of pups that must be harvested each year to maintain an equilibrium is well above those currently taken by commercial pup operations. Because an immediate reduction in the pup population does not affect pup production for at least six years, the time required to establish equilibrium may be longer than desirable. Harwood (1978) also calculated the resilience of equilibria for each kind of harvest. Resilience is measured by the characteristic return time (May, Conway, *et al.*, 1974), which is the speed with which the population returns to the neighborhood of an equilibrium after a reduction in the stock level. The return times to equilibria associated with proportional harvests were lowest near the MSY stock level, and they increased rapidly for equilibria associated with lower stock levels. With the exception of equilibria near MSY, return times for equilibria with either constant or proportional pup quotas were greater than the return time for the unexploited equilibrium. Thus, while proportional pup harvests can be used to define stable equilibria at relatively low stock levels, the difficulty of implementing large pup harvests and the low resilience of the resulting equilibria prompted Harwood (1978) to examine joint pup and adult harvesting.

Harwood (1978) noted that, because the population growth rate is sensitive to small changes in adult survival, a small adult harvest in conjunction with a pup harvest should stabilize the population at a relatively low stock level. Further, since female adults can be culled only during the breeding season, and the pups associated with culled adults will almost certainly die, an adult quota can be considered to take effect before any pups are born. Harwood (1978) also argued that if the adult quota is defined to be a proportion $0 \leq \gamma \leq 1$ of the previous year's pup production, then the population should be more rapidly stabilized by a joint pup-adult harvesting strategy. Thus he explored both fixed and proportional pup harvesting together with the adult harvesting strategy

$$h_6 = \gamma x_1.$$

In both cases there were combinations of pup and adult harvests that resulted in asymptotically stable stock levels well below the MSY level. The stabilizing effect of the adult quota resulted in the ability to choose a combination of pup and adult quotas for a desired equilibrium stock level that had a return time less than the return time of the unexploited equilibrium. Because of the destabilizing effect of the constant pup quota, combinations of pup and adult quotas that resulted in low return times could only be found for relatively low pup quotas.

In summary, the management goal of maintaining a stable and resilient grey seal stock below current levels can be achieved with a combination of adult and pup harvesting. The use of pup quotas alone is less desirable because of the potentially long return times associated with equilibria and because of the difficulty of implementing large proportional harvests. Although constant pup quotas are easier to implement, they are destabilizing when used without adult harvests. Combinations of small constant pup quotas and proportional adult harvests result in the desired equilibrium conditions.

6.2 MASS REARING INSECTS

6.2.1 Introduction

Demographic equations derived from static life tables appear to be unsuitable for modeling the growth of insect populations in the field. The primary problem is the high variability in survivorship caused by extensive fluctuations in daily mortality rates, as well as considerable spatial differences in these rates over distances of several meters. For example, a population of aphids in an alfalfa field may quickly succumb to an outbreak of a fungal epizootic that wipes more than 95% of the population. If one is unable to predict the occurrence of such events (which occur quite frequently during the rainy season) or estimate how many individuals survive the epizootic, then it is hopeless to try and predict the size of the population at some future date. Also, insects are poikilotherms. The rate at which they grow and the time taken to make the transition from one life stage to the next (e.g., egg, larva, pupa, adult) critically depend on temperature. For example, the different micro-climatic conditions presented by the north- and south-facing sides of the same tree affect the rate at which individual insects will grow and develop in these neighboring locations. Much work has been undertaken in trying to model the abundance of insect pests in agricultural systems. The models are more complicated than the framework presented in Chapters 2 and 3, because of the need to incorporate age-stage structure, temperature and other environmental factors, and nutritional factors into the model.

337

A framework for modeling insect populations in the field is based on the discretization of age-time von Foerster diffusion equations. These equations, without the elaboration of environmental and stage factors, represent a continuous time analog of the Leslie model. The elaborations to include environmental and nutritional factors, however, are needed to make the model more realistic (see Hughes, 1963; Sinko and Streifer, 1967; Hughes and Gilbert, 1968; Gutierrez *et al.*, 1981; and for a review see Getz and Gutierrez, 1982).

The constant-coefficient Leslie model has been applied with some success to insect laboratory cultures. The environmental conditions under which such populations grow and develop can be rigorously controlled. The populations are easily monitored and life-table data can be collected as often as required to obtain the level of resolution appropriate for a particular population model. The most natural time unit is the 24-hour day, although it may sometimes be expedient to select 2- or 3-day periods as a single unit of time.

Insects are reared for a number of reasons. Laboratory cultures may be used to provide a source of food for insectivorous animals that are themselves being studied in a laboratory setting. The most important commercial application of insect rearing, however, is in the area of biological control. Japan and Mexico, for example, have facilities to mass rear tens of millions of fruit fly (Tephritidae) per day for use in sterile insect release (SIR) programs. Fruit flies are major pests of agriculture in many countries (Mangel, 1985). California recently spent $100 million to eradicate a Mediterranean fruit fly (*Ceratitis capitata*) infestation, and SIR was a major component of that effort. Other examples of biological control programs include the release of ladybird beetles (Coccinellidae) which feed on aphids, predatory mites (Phytoseiidae) that have been selected for resistance to certain pesticides (primarily in almond orchards), and parasitoid wasps (*Trichogramma* spp.) whose larvae feed on the larvae of certain moths.

6.2.2   Fruit Fly Demography

The Mediterranean fruit fly infestation in California that began in 1980 prompted several demographic studies of this species. Carey (1982) made an initial cut at this task, using data obtained by Shoukry and Hafez (1979) who had reared this fly under a constant laboratory temperature of 25°C. They had collected interstage (egg, larva, pupa, adult) age-specific survival rates. These data were extrapolated by Carey so that they could be used in a model for which the basic time unit was 2 days. Thus in Table 6.1, the age-specific larval survival rates were obtained by Carey by interpolating so that the correct stage-specific rates were obtained. For example, an 80% pupal survival rate from day 12 to day 24 was obtained by interpolating between $l_{12} = 0.40$ and $l_{24} = 0.32$ (note that $l_{24}/l_{12} = 0.8$, which corresponds to an 80% pupal survival rate). Using these data and the formulae given in Chapter 2, Carey calculated a number of quantities, some of which are given in Table 6.2.

Carey used the life-table data to construct a Leslie model, as described in Chapter 2, and projected the dynamics of the population beginning with an initial infestation of pre-ovipositional adults. The evolution of the proportional stage structure of such a population through time is illustrated in Figure 6.4. Of course this analysis only applies to populations reared in the laboratory at 25°C (under the same nutritional conditions as the population from which the data were gathered), but indicates some important features that could apply in the field. The most striking feature is the initial oscillations in the stage structure. These oscillations suggest that control methods need to be timed at a point when the stage class most vulnerable to a particular type of control is relatively preva-

TABLE 6.1. Life history parameters for the Mediterranean fruit fly (Carey, 1982).

Stage	i	l_i	b_i	Stage	i	l_i	b_i
Egg	2	1.00	0.0	Female adult	38	0.26	26.5
				(continued)	40	0.25	26.0
Larva	4	0.83	0.0		42	0.24	25.7
	6	0.69	0.0		44	0.23	25.4
	8	0.58	0.0		46	0.22	24.6
	10	0.48	0.0		48	0.21	23.7
					50	0.20	22.6
Pupa	12	0.40	0.0		52	0.19	21.6
	14	0.39	0.0		54	0.18	20.2
	16	0.38	0.0		56	0.17	18.8
	18	0.37	0.0		58	0.16	17.4
	20	0.34	0.0		60	0.15	15.3
	22	0.32	0.0		62	0.14	13.7
					64	0.13	12.1
Female adult	24	0.32	0.0		66	0.11	10.4
(pre-oviposition)	26	0.31	0.0		68	0.10	9.1
					70	0.08	7.5
Female adult	28	0.31	2.4		72	0.06	6.4
(oviposition)	30	0.30	7.1		74	0.04	5.1
	32	0.29	16.5		76	0.02	4.0
	34	0.28	22.4		78	0.02	2.8
	36	0.27	25.7		80	0.02	1.2

lent. Thus, to a population following the structure depicted in Figure 6.4, one would apply a larvacide around day 20 or 65, but not around day 45 when only a small proportion of individuals is in the larval stage.

In a follow-up study, Carey (1984) reared the Mediterranean fruit fly in the laboratory on a number of different host fruits. A summary of the life table analysis for populations developing in five host fruits is given in Table 6.3. From these results it is apparent the mango is the superior host. In nectarine, although the larvae develop much more rapidly than in peach, significantly lower larval survival rates reduce the reproductive value of the population so that pear is a superior host as evidenced by the more than 10% faster doubling time. From a management point of view, however, larvacides would be more effective in

TABLE 6.2. Population parameters calculated from the Mediterranean fruit fly life-history data in Table 6.1.

Parameters	Values
Stable stage structure	$(0.298, 0.497, 0.156, 0.018, 0.031)^a$
Net reproductive value $(R_0)^b$	84.1 female eggs per female
Mean generation time $(T)^c$	40.8 days
Rate of increase $(\lambda_\Delta)^d$	1.115
Doubling time[e]	6.38 days

[a] Proportion of the population respectively in the egg, larval, pupal, preovipositional, and ovipositional stage classes.

[b] See equation (2.8).

[c] See equation (2.9).

[d] This should be the Perron root discussed in Section 2.2.5, but it appears that Carey has evaluated it corresponding to a daily time step rather than the 2-day time step set up in Table 6.1.

[e] This is the number of days that the population takes to double based on the geometric growth rate given by the Perron root (see note d above).

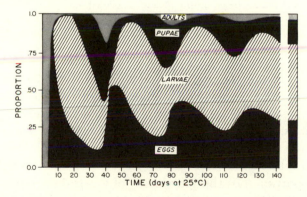

FIGURE 6.4. Age structure of Mediterranean fruit fly projected from an initial infestation of pre-ovipositional female adults. (Data applies to insects developing under a constant temperature of 25°C—reproduced with permission (Carey, 1982).)

peach and pear than nectarine, fig, or mango, because of the relatively greater proportion of the population in the larval stage in peach and pear.

341

TABLE 6.3. Parameters from an analysis of life-table data collected from five laboratory populations reared at 25°C (Carey, 1984).

Host	Parameter				Stable stage distribution (mean development time)[a]			
	$\lambda_{\triangle}$[b]	R_0[c]	τ[d]	DT[e]	Eggs	Larvae	Pupae	Adults
Mango	1.178	444	35.8	4.2	28	49(7.2)	20(11.5)	3
Fig	1.152	250	37.4	4.9	32	51(8.2)	14(11.9)	3
Peach	1.144	301	40.6	5.1	27	59(11.1)	11(11.9)	3
Nectarine	1.129	125	38.0	5.7	37	49(8.0)	11(12.5)	3
Pear	1.128	250	44.0	5.8	27	64(13.9)	8(12.2)	2

[a] Proportion of the population respectively in the egg, larval, pupal, and adult female stage classes. Since the time spent in each stage class is no longer the same as that indicated in Table 6.1, the mean development time in the larval and pupal stage classes is indicated in parentheses.

[b] This is the Perron root (see Section 2.2.5) corresponding to a daily time step.

[c] See equation (2.8).

[d] See equation (2.9).

[e] Doubling time—this is the number of days that the population takes to double based on the geometric growth rate given by the Perron root.

6.2.3 Mass Rearing Fruit Fly

The reasons for rearing fruit fly on a large scale have already been discussed. The flies are reared in a factory-type facility that contains a self-sustaining population. From time to time new genetic stocks may be introduced into the population to ensure its continued viability. At given intervals of time (every second day, for example), a proportion h of the individuals of a certain age i_h, say, in the larval stage are removed from the population. Furthermore, since adults become less fecund with age, it may be efficient to remove individuals at a particular discard age i_δ, rather than wait for individuals to die from natural causes.

In a mass-rearing facility, the population is reared under controlled highly favorable environmental and dietary conditions so that density-mediated population regulation

342

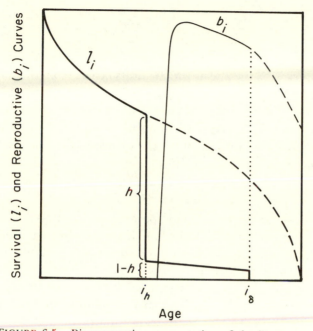

FIGURE 6.5. Diagrammatic representation of the insect mass-rearing problem. (Adapted from Carey and Vargas, 1985.)

may not play a role (at least at the densities under which the insects are reared). In this case the life-table analyses of Chapter 2 and the previous subsection apply. As we see from Table 6.4, such populations reproduce prolifically, and have a population doubling time of around 5 days. For the rearing facility population to be in equilibrium, the harvesting variable h and discard age i_δ must be chosen such that the net reproductive value of the managed population satisfies

$$R_0(h, i_\delta) = 1. \tag{6.5}$$

Recalling equation (2.8), where R_0 is expressed for a natural population, it follows that equation (6.5) represents

343

the constraint

$$\sum_{i=1}^{i_h-1} b_i l_i + (1 - h) \sum_{i=i_h}^{i_\delta} b_i l_i = 1. \qquad (6.6)$$

Suppose i_a is the age at which adults eclose from the pupal stage (thus as depicted in Figure 6.5, $i_h < i_a < i_\delta$). Since $h(l_h/l_0)$ is the number of pupae harvested per egg laid (l_0) and $(1 - h) \sum_{i=i_a}^{i_\delta} /l_0$ is the number of adults per egg laid, it follows that $Y(h, i_\delta)$, the yield of pupae per maintained adult, is

$$Y(h, i_\delta) = \frac{h l_h}{(1 - h) \sum_{i=i_a}^{i_\delta} l_i}. \qquad (6.7)$$

Note that if the life-table parameters apply only to females, then Y has the units female pupae per female adult and, assuming a 50:50 sex ratio, must be multiplied by 2 to get the total number of pupae per female adult. Since $b_i = 0$ for all $i \leq i_a$, equation (6.6) can be used to express h in terms of $\sum_{i=i_a}^{i_\delta} b_i l_i$, which when substituted in equation (6.7) results in an expression for $Y(h, i_\delta)$ that depends solely on the life-table parameters l_i and b_i, the harvesting age parameter i_h (which is determined from biological studies on when it is best to irradiate the pupae), and the discard age variable i_δ. Thus in equation (6.7), i_δ is the only parameter that can be manipulated to maximize Y, the number of pupae per adult. If adults are relatively expensive to maintain then maximizing $Y(h(i_\delta), i_\delta)$ as a function of i_δ is an important consideration in the production process.

Carey and Vargas (1985) evaluated $Y(h(i_\delta), i_\delta)$ for three species of fruit fly. The results they obtained are illustrated in Figure 6.6. They also calculated a number of other quantities including the stable stage distribution listed in Table 6.4. The proportion of melon fly adults in the stable stage distribution is almost twice as large as for the other two fruit fly. Further, from Figure 6.6 we

344

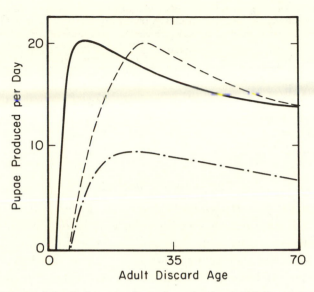

FIGURE 6.6. Daily rate of pupal production in Mediterranean fruit fly (——), Oriental fruit fly (— — —), and melon fly (— · —). (Adapted from Carey and Vargas, 1985.)

see that the maximum pupal (both sexes) production rate per adult female for the melon fly is less than half that of the other two species. Thus the number of pupae that a particular facility can produce depends very much on the species being cultured.

The analysis of fruit fly production was taken a step further by Plant (1986) who considered the problem of how to manage the population from starting up production until the facility had reached its maximum rearing capacity. He also developed a model for the interaction between the wild and introduced sterile populations. These are important problems to analyze but, at the equilibrium mass-rearing level, an important problem remains which will be dealt with in the next subsection.

345

TABLE 6.4. Percentage of individuals in each life stage exclusive of harvested pupae, for three species of fruit fly (Carey and Vargas, 1985).

Species	Discard age (δ)	Percentage of total population			
		Eggs	Larvae	Pupae	Adults
Mediterranean fruit fly, *Ceratitis capitata*	14	12.1	40.2	47.2	0.5
Oriental fruit fly, *Dacus dorsalis*	20	13.6	41.4	44.5	0.5
Melon fly, *D. cocurbitae*	25	17.0	42.6	49.5	0.9

6.2.4 An Age-Stage Model

In many insect populations reared under strictly controlled laboratory conditions there is sufficient genetic variability so that not all individuals enter or leave a particular stage class at the same age. In fact, the degree of overlap can be quite large as is evidenced from Figure 6.7, which represents a diamondback moth (*Pultella xylostella* [L.]) reared in the laboratory at a constant temperature of 25°C. The larval and pupal stages, for example, overlap from day 11 to 16.

To be able to account for this overlap, Chi and Liu (1985) proposed the following extension to the classical life-table analysis, discussed in Chapter 2. They defined variables x_{ij} to represent the number of individuals of age i in stage class j. They then replaced the age-transition equation (2.3) with an age-stage transition equation

$$x_{ij}(t + 1) = a_{i-1\,j}x_{i-1\,j}(t) + c_{i-1\,j-1}x_{i-1\,j-1}(t), \qquad (6.8)$$

where a_{ij} and c_{ij} respectively represent the proportion of individuals that make the transition within and between stage classes from one age class to the next. This equation holds for the age classes $i = 1, \ldots, n - 1$, where it is assumed that no individuals survive beyond age n (an extra term, as in equation (2.4), can be added if individuals

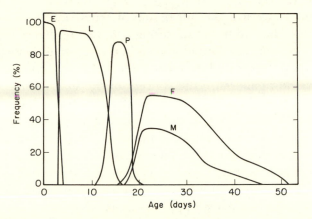

FIGURE 6.7. Stage frequency development pattern of diamond-back moth reared from newly laid eggs at 25°C: E, egg; L, larvae; P, pupae; F, females; M, males. (Adapted from Chi and Liu, 1985.)

are assumed to survive beyond age n). The equation also holds for all j, except $j = 1$ and $j = m$. When $j = 1$ (egg-stage class), there is no stage class below (that is, $b_{i0} \equiv 0$). The last two stage classes, $m - 1$ and m, are the female and male class, respectively. Since males originate from the pupal stage class $m - 2$, equation (6.8) is modified for $j = m$ to read

$$x_{im}(t + 1) = a_{i-1\,m}x_{i-1\,m}(t) + c_{i-1\,m-2}x_{i-1\,m-2}(t). \quad (6.9)$$

Note that if the sexes are split in the model at an earlier life stage (for example, differences between male and female larvae might become apparent at a particular instar), then the equations can be modified accordingly. Finally, eggs are produced only by the female life stage $j = m - 1$. Thus if i_f is the first age at which a female becomes fecund, then

$$x_{01}(t + 1) = \sum_{i=i_f}^{n} b_i x_{i\,m-1}.$$

347

The variable x_{01} is our "aggregated" variable for this model (since it is the linear sum of other state variables), but has the additional subscript 1 to denote that it is also the initial number of individuals in the egg-stage class (this notation differs slightly from Chi and Liu, 1985).

We now generalize the concept of the survivorship column l_i introduced in Section 2.1.2 to include stage structure as follows. Set l_{01} to an appropriate value. (Here, following our convention in the earlier chapters, we select $l_{01} = 1$ so that l_{ij} has the interpretation as the proportion of individuals that survive from one age-stage class to the next). Then generate entries l_{ij} using equations (6.8) and (6.9) with l_{ij} in place of $x_{ij}(t)$; that is,

$$l_{ij} = a_{i-1\,j}l_{i-1\,j} + c_{i-1\,j-1}l_{i-1\,j-1}, \qquad (6.10)$$

and so on. Then clearly l_{ij}/l_{01} represents the proportion of individuals in the jth stage class that survive to age i, and

$$l_i = \sum_{j=1}^{m} l_{ij},$$

when divided by l_0, represents the proportion of individuals that survive to age i: our classical life-table survivorship column. Thus all the usual life-table analyses follow, including calculation of the Perron root λ_Δ from the equation (cf. equation (2.35))

$$\sum_{i=1}^{n} \frac{b_i \sum_{j=1}^{m} l_{ij}}{\lambda^i}.$$

The stable age-stage distribution can be generated from the equation (cf. equation (2.14))

$$\lambda_\Delta x_{ij} = a_{i-1\,j}x_{i-1\,j} + c_{i-1\,j-1}x_{i-1\,j-1},$$

and so on; the stable age or stage distributions can be derived by respectively summing over the indices j or i as indicated in the matrix representation of the state variables illustrated in Figure 6.8.

Age–Stage Representation

E	L	P	F	M	Age

$$
\begin{bmatrix}
x_{11} & 0 & 0 & 0 & 0 \\
x_{21} & x_{22} & 0 & 0 & 0 \\
x_{31} & x_{32} & 0 & 0 & 0 \\
0 & x_{42} & x_{43} & 0 & 0 \\
0 & x_{52} & x_{53} & 0 & 0 \\
0 & 0 & x_{63} & x_{64} & x_{65} \\
0 & 0 & x_{73} & x_{74} & x_{75} \\
0 & 0 & 0 & x_{84} & x_{85} \\
0 & 0 & 0 & x_{94} & x_{95} \\
\vdots & \vdots & \vdots & \vdots & \vdots \\
0 & 0 & 0 & x_{35\,4} & x_{35\,5}
\end{bmatrix}
\begin{bmatrix}
x_{11} \\
\Sigma_{j=1}^{2}\, x_{2j} \\
\Sigma_{j=1}^{2}\, x_{3j} \\
\Sigma_{j=2}^{3}\, x_{4j} \\
\Sigma_{j=2}^{3}\, x_{5j} \\
\Sigma_{j=3}^{5}\, x_{6j} \\
\Sigma_{j=3}^{5}\, x_{7j} \\
\Sigma_{j=4}^{5}\, x_{8j} \\
\Sigma_{j=4}^{5}\, x_{9j} \\
\vdots \\
\Sigma_{j=4}^{5}\, x_{35\,j}
\end{bmatrix}
$$

$$
\begin{bmatrix}
\displaystyle\sum_{i=1}^{3} x_{i1} & \displaystyle\sum_{i=2}^{5} x_{i2} & \displaystyle\sum_{i=4}^{7} x_{i3} & \displaystyle\sum_{i=6}^{35} x_{i4} & \displaystyle\sum_{i=6}^{35} x_{i5}
\end{bmatrix}
$$

Stage

FIGURE 6.8. An age-stage representation of the state of an insect population where individuals are in the egg stage (E) for 1 to 3 days, in the larval stage (L) from days 2 to 5, the pupal stage (P) from days 4 to 7, and female (F) and male (M) adult stages from days 6 to 35. (Adapted from Chi and Getz, 1988.)

6.2.5 Mass Rearing the Potato Tuberworm

The potato tuberworm, *Phthorimaea operculella* (Zeller), is cultured in laboratories as a food source for such predatory insects as coccinellids and lacewings, which themselves are reared for release as biological control agents. Data from a particular potato tuberworm culture maintained at the Division of Biological Control, University of California at Berkeley, was collected by Chi (1988) to ob-

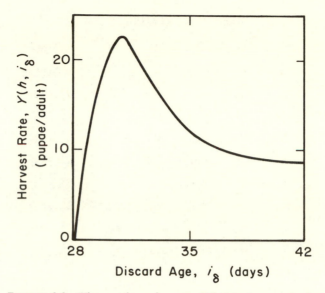

FIGURE 6.9. The number of pupae, $Y(h, i_\delta)$, that can be harvested per adult plotted as a function of the discard age i_δ for an equilibrium potato tuberworm laboratory culture.

tain the survivorship and natality parameters listed in Table 6.5. These data were then used by Chi and Getz (1988) to calculate some growth parameters and stable age and stage distributions associated with this population, as well as the optimal discard age and corresponding proportion of pupae harvested to maintain an equilibrium population level. Some results from these calculations are listed in Table 6.6, while a plot of $Y(h, i_\delta)$, the number of pupae harvested per adult (see equation (6.7)), is given in Figure 6.9. From this figure we see that the optimal discard age is 31 days, with corresponding yield of $Y(h, i_\delta) = 22.9$. If adults are removed only when dead, the yield falls to 8.7 pupae per adult. Since the mean longevity of adults is relatively short (8.3 days) and adults do not require feeding, there may not be an advantage to maintaining the population under an optimal discard-age policy.

6.3 OVERVIEW

6.3.1 Importance of Theory

How many of us would fly in an airplane built in someone's backyard without any blueprints? This airplane may *look* like others we have seen fly, but that is not enough to ensure that it will fly. Someone needs to calculate the maximum load that the wings can bear to be sure that they won't fall off during flight. It is wise to demand that airplanes be built from blueprints designed by an engineer. Should we demand any less of ourselves than to manage resources according to policies designed by resource scientists? Granted that there is much more uncertainty associated with calculating the growth rate of a population than there is with the forces that act on an airplane wing; but the absence of theory in resource management is just as likely to lead to a disaster, as is the absence of theory in aviation.

Resource management is a melding of population biology and economics, both of which have less well-developed theories than the mechanics of flight. The simplest theories in resource management are based on single-variable (lumped) population models that, typically, predict the total biomass of the resource at different points in time. This approach leads to a good understanding of the qualitative differences in the stability of a resource subject to constant-yield versus proportional-harvesting policies, or the economics of resources exploited by a single individual versus a group of competing inviduals (for example, see Clark, 1976, 1985). A number of central questions in resource management, however, pertain to the age or size structure of the resource. It may be possible to utilize small fish by grinding them up for fish meal or pressing

351

TABLE 6.5. Survivorship and fecundity columns for an age-stage structured representation of a potato tuberworm population at 25°C (see Chi, 1988).

Age	Survivorship					Fecundity (Female)
	Eggs	Larvae	Pupae	Female	Male	
1	1.00	—	—	—	—	—
2	1.00	—	—	—	—	—
3	1.00	—	—	—	—	—
4	1.00	—	—	—	—	—
5	1.00	—	—	—	—	—
6	1.00	—	—	—	—	—
7	—	1.00	—	—	—	—
8	—	1.00	—	—	—	—
9	—	1.00	—	—	—	—
10	—	1.00	—	—	—	—
11	—	1.00	—	—	—	—
12	—	1.00	—	—	—	—
13	—	1.00	—	—	—	—
14	—	1.00	—	—	—	—
15	—	1.00	—	—	—	—
16	—	1.00	—	—	—	—
17	—	0.90	0.05	—	—	—
18	—	0.75	0.20	—	—	—
19	—	0.40	0.55	—	—	—
20	—	0.10	0.85	—	—	—
21	—	—	0.95	—	—	—
22	—	—	0.95	—	—	—
23	—	—	0.95	—	—	—
24	—	—	0.95	—	—	—
25	—	—	0.95	—	—	—
26	—	—	0.95	—	—	—
27	—	—	0.90	0.05	—	0.0
28	—	—	0.75	0.15	—	0.0
29	—	—	0.70	0.20	—	36.5
30	—	—	0.35	0.35	0.20	39.7
31	—	—	0.10	0.45	0.35	48.9
32	—	—	—	0.55	0.35	14.4
33	—	—	—	0.55	0.35	23.5
34	—	—	—	0.55	0.35	6.2
35	—	—	—	0.55	0.30	3.2
36	—	—	—	0.50	0.20	0.4
37	—	—	—	0.50	0.15	0.3
38	—	—	—	0.30	0.05	0.7
39	—	—	—	0.20	—	0.0
40	—	—	—	0.10	—	0.0
41	—	—	—	0.10	—	0.0
42	—	—	—	0.05	—	0.0
43	—	—	—	0.05	—	0.0

TABLE 6.6. Population parameters and stable stage distribution for the potato tuberworm with and without harvesting (see Chi and Getz, 1988).

Harvesting	Population parameters			Stable stage distribution			
	$\lambda_\Delta{}^a$	$R_0{}^b$	T^c	Egg	Larvae	Pupae	Adult
Without	1.146	69.7	31.2	56.5	36.1	6.2	1.2
With	1	1	—	32.6	66.0	0.8	0.6

^{a} Perron root—see equation (6.10). Units are days^{-1}.

^{b} Reproductive value—see equation (2.8). Units are offspring per individual.

^{c} Mean generation time—see equation (2.9). Units are days. The mean generation time with harvesting was not calculated in Chi and Getz (1988).

them for fish oil, but only the large fillets can be sold to restaurants. Only trees of a certain size can be sawed into large planks, and trophy hunters are interested only in male deer with large antlers.

The instantaneous growth rate of age- and stage-structured populations depends on the density of individuals in the populations and on the proportion of reproductively mature individuals (for example, see the fir-tree growth models in Chapter 4 and the elephant case studies in Chapter 6). There are many economically important resources based on structured populations, as indicated by the selected fisheries, forest, wildlife, and pest management examples presented in this book. A density-independent theory for modeling such populations goes back to the 1940s, but density-independent models are inadequate for modeling populations that exibit strong density-dependent fecundity and survival of newborns. A density-dependent theory for modeling age-structure populations has blossomed only over the last ten years, primarily in fisheries where the nonlinear recruitment process has been linked to linear survival and transition. These models provide the basis for developing the stage-structured aggregated density-dependent variable model that is the umbrella for all the case studies presented in Chapters 4 to 6.

The economic side of the theory of resource management relates to the profit that can be made from exploiting a population. In this book we deal only with relatively simple yield- and rent-maximization problems; in some cases finding the optimal harvesting policy is subject to constraints that minimize the variation in annual yield or biological stock levels, or even ensure that stock levels never fall below some prescribed minimum value. The solution procedures are obtained from *optimization theory*, which is a well-developed branch of mathematics. Thus the difficulties in applying optimization theory to resource management are generally not related to the solution procedures, but to whether we have adequate population models and appropriately defined problems. The question of adequate models is a difficult one. It is not really addressed in this book, except for our comparative discussion in Section 5.2.7 of stage-structured versus single-tree simulators for forest stand management. However, we do address questions relating to the formulation of management problems, especially the issue of planning horizons and conversion policies in forest stand management (Section 5.3).

6.3.2 Applying the Theory

The models developed in this book can be applied to resource harvesting problems only if life-table parameters and density-dependent relationships have been quantified. How to obtain such data is a major problem, one that is dealt with by field biologists in each area of application. These are not questions that we have dealt with in this book, but it is very important to keep in the back of our minds that the success or failure in applying models to particular resource management problems is ultimately determined by the quality of our biological data.

The examples in Chapters 3 to 6, drawn from the fisheries, forestry, wildlife, and pest management literature, should convince the reader of the general applicability of the theory developed in Chapters 2 and 3 to problems in biological resource management. Of course we have not covered the full range of applications in the literature since, to a large extent, we have focused only on areas of research that we are familiar with in fisheries, forestry, and pest management. Most of the applications we present are developed in the literature in the framework of theory presented in Chapters 2 and 3, although the original notation may be different. In the grey seal harvesting problem discussed in Section 6.1.3, however, we molded the approach taken in the original study into our more general framework.

Each of the applications demonstrates a particular aspect of the general aggregated-nonlinear stage-structured modeling framework. In fisheries analysis, we primarily use age-structured models with nonlinear recruitment that depends on a single aggregated stock-biomass variable. The problem of managing fisheries under both deterministic and stochastic conditions is discussed. In forest-stand management analyses we employ stage-structured models, but use only deterministic techniques. In the mammal and insect mass-rearing examples we apply both the straightforward age-structure theory and extensions that include different types of stage classes (pregnancy states in the case of elephants; and egg, larval, pupal, and adult life stages in the case of insects). All these examples serve to illustrate the versatility and flexibility of the modeling framework developed in Chapters 2 and 3.

We have deliberately restricted the theory part of the book, however, to an analysis of stage-structured models that only allow for single-step transitions in one direction

355

along an ordered sequence of stage classes. This was done primarily because this class of models has broad applicability to fishery and forestry management problems, and secondly because this is the first time the theory of such models has been presented in a comprehensive setting. More complicated transition models have been applied, especially to problems involving separation of both sexes or organisms that have multistate life histories (for example, see Caswell and Weeks, 1986, and Section 6.3.4). Many of these more detailed transition models avoid the complications of density-dependent parameters, as is the case with elephant population models developed by Wu and Botkin (1980) (see matrix representation (6.1) and (6.2)). It remains to be seen how easily a more comprehensive nonlinear transition theory can be developed.

6.3.3 *Future Directions*

The extension of theory to include transitions between any two stage classes poses some difficulties, in general, because the progression/survivorship matrix G can no longer be inverted explicitly, as was done in Lemma 2.3. Some forms of G, however, are easier to work with than others, and the question of which forms can be explicitly inverted deserves some attention. A general matrix G is no more difficult to use in simulation studies, however, than the specific matrix G introduced in Chapter 2.

The most difficult aspect of developing general theories is to fully embrace the dynamic and stochastic character of resource management problems. It is evident from Chapter 3 that one can make considerable progress analyzing deterministic systems under equilibrium conditions (for example, Theorem 3.2), but to say something about dynamic, let alone stochastic, systems (for example, Theorem 3.3) is much more difficult. Deterministic and equilibrium analyses do provide frames of reference for addressing stochastic dynamic problems, but for many resource systems stochastic analyses will lead to a much better understanding of the management problem. The fisheries case studies discussed in Section 4.5 provide examples of the value of a stochastic approach. However, much more work needs to be undertaken in developing techniques for managing resources under stochastic conditions, especially relating to the use of particular measures (e.g., the proportion of large fish in the catch). In stochastic resource management, constant monitoring of the resource is essential, as only timely corrective measures can ultimately save a resource from collapse. Such "adaptive" resource management has been explored, primarily using lumped (single-variable) models (see Walters, 1986). Using age-structured analyses, however, one can monitor the relative proportion of individuals in various age classes and use this additional information to gauge the health of the stock and adjust harvest rates.

Lumped population models are more easily analyzed in a stochastic dynamic setting than structured population models, because multidimensional problems are often analytically intractable, and the difficulties associated with numerical computations increase as a power of the dimension of the problem. Thus scalar models have a strong advantage over the structured models discussed here, in situations where there is insufficient data to estimate the several (or even many) parameters associated with structured models. It is apparent from the lumping process, discussed in Section 3.6, that a number of stringent assumptions need to be made with regard to age-specific fecundity, mortality, and vulnerability to harvesting. An important area of research is to assess how much accuracy we lose in predicting yield and stock levels with lumped models. To this, however, we need to compare predictions and solutions obtained using both structured and lumped models. Also, it may transpose that such critical questions as discarding in multispecies fisheries (see Section 4.7.3) can be adequately addressed only using age-structured models.

Forest management analyses have traditionally been subject to more rigorous application of mathematical optimization techniques than other resource systems (for example, see Buongiorno and Gilless, 1987). The reason is that forest stands behave in a more predictable manner than animal populations over the time scale of several years, and trees are also more easily observed and hence counted and measured. This allows for greater reliability in making economic assessments, which in turn lends itself towards more sophisticated methods of analysis. In comparing the stage-structured forest-stand management formulations developed in Chapter 5 with single-tree stand-growth simulators, as discussed in Section 5.5, it is apparent that techniques in optimization theory and systems analysis are far more readily implemented using the stage-structured approach. Furthermore, stochastic forest-stand management is also facilitated by the simpler form of stage-structured models. For stochastic problems, the linearized models discussed in Section 3.3.3 might prove extremely useful in applying stochastic linear theory (cf. Goodwin and Sin, 1984) to resource management problems. This would avoid having to resort to Monte Carlo methods, as used in the fisheries problems discussed in Section 4.5. Linearized methods are more appropriately applied to forest-stand than fisheries management problems (for the same reasons as mentioned above).

The models presented in this book all assume that the populations are insulated from the dynamics of other populations; that is, we have removed the ecological aspects of the problem. Even in the multispecies fisheries in Section 4.7 we assumed that links between species were through fishing technology and not ecology. Very little work has been undertaken on the analysis of interactions between age- or stage-structured populations. Travis *et al.* (1980)

have analyzed competing age-structured populations subject to Beverton and Holt-like (compensatory) recruitment
processes. This analysis was taken further by Bergh and
Getz (in press b) in a fisheries management context. The
whole area of interacting age-structured populations is
ripe for research, including the analysis of prey-predator
systems.

6.4   CONCLUSION

In this book, we have packaged population harvesting
theory (Chapters 2 and 3) with applications in fisheries,
forest, wildlife, and pest management (Chapters 4, 5, and
6) under a unified notation. We hope that this will help
researchers in various areas of resource management to
appreciate and learn from the resource problems outside
their immediate sphere of interest; that is, we hope that
this book, and the notation we use, will facilitate a cross-
fertilization of ideas and make transparent the relationship between problems formulated in different applied areas. If we can get forestry managers to read Chapters 4
and 6 and fisheries managers to read Chapters 5 and 6, we
will have gone a long way towards justifying our efforts in
writing this book.

References

Adams, D. M., 1976. A note on the interdependence of stand structure and best stocking in a selection forest. Forest Sci. 22:180–184.

Adams, D. M., and A. R. Ek, 1974. Optimizing the management of uneven-aged forest stands. Can. J. For. Res. 4:274–287.

Allard, J., D. Errico, and W. J. Reed, 1986. Irreversible investment and optimal forest exploitation. Tech. Report No. 38, Dept. of Statistics, University of British Columbia.

Archibald, C. P., D. Fournier, and B. M. Leaman, 1983. Reconstruction of stock history and development of rehabilitation strategies for Pacific ocean perch in Queen Charlotte Sound, Canada. N. Am. J. Fish. Manage. 3:283–294.

Arney, J. D., 1985. A modeling strategy for the growth projection of managed stands. Can. J. For. Res. 15:511–518.

Bailey, K. M., R. C. Francis, and P. M. Stevens, 1982. The early life history of Pacific hake (*Merluccius productus*). Calif. Coop. Oceanic Fish. Invest. Rep. 23:81–98.

Baranov, F. I., 1925. On the question of the dynamics of the fishing industry. Byull. Rybn. Khoz. 8:7–11. (Translated from Russian by W. E. Ricker, 1945.)

Bare, B. B., and D. Opalach, 1987. Optimizing species composition in uneven-aged forest stands. Forest Sci. 33:958–970.

Bazaraa, M. S., and C. M. Shetty, 1979. Nonlinear Programming: Theory and Algorithms. Wiley, New York.

Beddington, J. R., and R. M. May, 1977. Harvesting natural populations in a randomly fluctuating environment. Science 197:463–465.

361

REFERENCES

Beddington, J. R., and D. B. Taylor, 1973. Optimum age specific harvesting of a population. Biometrics 29:801–809.

Belcher, D. W., M. R. Holdaway, and G. J. Brand, 1982. A description of STEMS: the stand entry evaluation and modeling system. USDA For. Serv. Gen. Tech. Rep. NC-79, North Central Forest Exp. Stn., St. Paul, Minn.

Berck, P., and T. Bible, 1984. Solving and interpreting large-scale harvest scheduling problems by duality and decomposition. Forest Sci. 30:173–182.

Bergh, M. O., 1986. The value of catch statistics and records of guano harvests for managing certain South African fisheries. Ph.D. thesis, University of Cape Town, Rondebosch, South Africa.

Bergh, M. O., and D. S. Butterworth, 1987. Towards rational harvesting of South African Anchovy considering survey imprecision and recruitment variability. In: A.I.L. Payne, J. A. Gulland, and K. H. Brink (eds.), The Benguela and Comparable Ecosystems. Special Publication, South African J. Marine Science 5:937–951.

Bergh, M. O., and W. M. Getz, in press a. Stability of discrete age structured and aggregated delay-difference population models. J. Math. Biol.

Bergh, M. O., and W. M. Getz, in press b. Stability and harvesting of competing populations with genetic variation in life history strategies. Theoret. Pop. Biol.

Best, E., 1961. Saving gear studies on Pacific Coast flatfish. Pacific Mar. Fish. Comm. Bull. 5:25–48.

Beverton, R.J.H., and S. J. Holt, 1957. On the dynamics of exploited fish populations. Ministry of Agriculture, Fisheries and Food (London), Fish. Invest. Ser. 2(19).

Biging, G. S., 1984. Taper equations for second-growth mixed conifers of northern California. Forest Sci. 30:1103–1117.

REFERENCES

Bosch, C. A., 1971. Redwoods: A population model. Science 172:345–349.

Botkin, D. B., J. M. Mellilo, and L. S.-Y. Wu, 1981. How ecosystem processes are linked to large mammal population dynamics. In: C. W. Fowler and T. D. Smith (eds.), Dynamics of Large Mammal Populations, pp. 373–387. Wiley, New York.

Brock, W. A., and J. S. Scheinkman, 1976. Global asymptotic stability of optimal control systems, with applications to the theory of economic growth. J. Economic Theory 12:164–194.

Brodie, J. D., and C. Kao, 1979. Optimizing thinning in Douglas fir with three-descriptor dynamic programming to account for accelerated diameter growth. Forest Sci. 25:665–672.

Brodie, J. D., D. M. Adams, and C. Kao, 1978. Analysis of economic impacts on thinning and rotation for Douglas fir using dynamic programming. Forest Sci. 24:513–522.

Bullard, S. H., H. D. Sherali, and W. D. Klemperer, 1985. Estimating optimal thinning and rotation for mixed-species timber stands using a random search algorithm. Forest Sci. 31:301–314.

Buongiorno, J., and J. K. Gilless, 1987. Forest management and economics. Macmillan, New York.

Buongiorno, J., and B. R. Michie, 1980. A matrix model of uneven-aged forest management. Forest Sci. 26:609–625.

Butterworth, D. S., D. C. Duffy, T. B. Best, and M. O. Bergh, 1988. On the scientific bases for reducing the South African seal population. South African J. Sci. 84:179–188.

Carey, J. R., 1982. Demography and population dynamics of the Mediterranean fruit fly. Ecological Modelling 16:125–150.

Carey, J. R., 1984. Host-specific demography of the

363

Mediterranean fruit fly, *Ceratitis capitata*. Ecological Entomology 9:261–270.

Carey, J. R., and R. I. Vargas, 1985. Demographic analysis of insect mass rearing: A case study of three tephritids. J. Econ. Entomol. 78:523–527.

Caswell, H. and D. Weeks, 1986. Two-sex models: Chaos, extinction and other dynamic consequences of sex. Am. Natur. 128:707–735.

Cawrse, D. C., D. R. Betters, and B. M. Kent, 1984. A variational solution technique for determining optimal thinning and rotational schedules. Forest Sci. 30:793–802.

Chang, S. J., 1981. Determination of the optimal growing stock and cutting cycle for an uneven-aged stand. Forest Sci. 27:739–744.

Chang, S. J., 1984. Determination of the optimal rotation age—a theoretical analysis. Forest Ecol. and Manage. 8:137–147.

Charlesworth, B., 1980. Evolution in Age-structured Populations. Cambridge University Press, Cambridge.

Chi, H., 1988. Life table analysis incorporating both sexes and variable development rates among individuals. Environ. Ent. 17:26–34.

Chi, H., and W. M. Getz, 1988. Mass rearing and harvesting based on an age-stage, two-sex life table: a potato tuberworm (Lepidoptera: Gelechiidae) case study. Environ. Ent. 17:18–25.

Chi, H., and H. Liu, 1985. Two new methods for the study of insect population ecology. Bull. Inst. Zool., Academia Sinica 24:225–240.

Clark, C. W., 1976. Mathematical Bioeconomics: The Optimal Management of Renewable Resources. Wiley-Interscience, New York.

Clark, C. W., 1985. Bioeconomic Modeling and Fisheries Management. Wiley-Interscience, New York.

REFERENCES

Cohen, J. E., 1976. Ergodicity of age structure in populations with Markovian vital rates. I: Countable states. J. Am. Stat. Assoc. 71:335–339.

Cohen, J. E., 1977a. Ergodicity of age structure in populations with Markovian vital rates. II: General states. Adv. Appl. Probab. 9:18–37.

Cohen, J. E., 1977b. Ergodicity of age structure in populations with Markovian vital rates. III: Finite state moments and growth rates: illustration. Adv. Appl. Probab. 9:462–475.

Cohen, J. E., 1979. Comparative statics and stochastic dynamics of age structured populations. Theor. Pop. Biol. 16:159–171.

Cohen, J. E., S. W. Christensen, and C. P. Goodyear, 1983. A stochastic age-structured populations model of striped bass (*Morone saxatilis*) in the Potomac River. Can. J. Fish. Aq. Sci. 40:2170–2183.

Cole, D. M., and A. R. Stage, 1972. Estimating future diameters of lodgepole pine. USDA Forest Serv. Res. Paper INT-131, Intermt. For. and Range Exp. Stn., Ogden, Utah.

Croze, H., A.K.K. Hillman, and E. M. Lang, 1981. Elephants and their habitats: how do they tolerate each other. In: C. W. Fowler and T. D. Smith (eds.), Dynamics of Large Mammal Populations, pp. 297–316. Wiley, New York.

Cushing, D. H., 1981. Fisheries Biology: A Study in Population Dynamics. University of Wisconsin Press, Madison.

Davis, L. S. and K. N. Johnson, 1987. Forest Management, 3d edition. McGraw Hill, New York.

Demetrius, L., 1971. Primitivity conditions for growth matrices. Math. Biosci. 12:53–58.

Demetrius, L., 1972. On an infinite population matrix. Math. Biosci. 13:133–137.

Doubleday, W. G., 1975. Harvesting in matrix populations. Biometrics 31:189–200.

Dreyfus, S. E., and A. M. Law, 1977. The Art and Theory of Dynamic Programming. Academic Press, New York.

Duerr, W. A., and W. E. Bond, 1952. Optimum stocking of a selection forest. J. Forestry 50:12–16.

Ek, A. R., 1974. Nonlinear models for stand table projection in northern hardwood stands. Can. J. For. Res. 4:23–27.

Emlen, J. M., 1984. Population Biology: The Coevolution of Population Dynamics and Behavior. Macmillan, New York.

Faustmann, M., 1849. Calculation of the value which forest land and immature stands possess for timber growing. In: Martin Faustmann and the Evolution of Discounted Cash Flow, pp. 18–34. Commonwealth For. Inst. Paper 42, Oxford, 1968 (translated by W. Linnard).

Ferguson, D. E., A. R. Stage, and R. J. Boyd, 1986. Predicting regeneration in the grandfir-cedar-hemlock ecosystem of the northern Rocky Mountains. Forest Science Monograph 26.

Fisher, A. C., 1981. Resource and Environmental Economics. Cambridge University Press, Cambridge.

Fisher, R. A., 1930. The Genetical Theory of Natural Selection. Clarendon Press, Oxford (reprinted and revised, 1958, Dover Publications, New York).

Forrester, J. W., 1961. Industrial Dynamics. MIT Press, Cambridge, Mass.

Fournier, D. A., and C. P. Archibald, 1982. A general theory for analyzing catch at age data. Can. J. Fish. Aq. Sci. 39:1195–1207.

Fowler, C. W., and T. D. Smith, 1973. Characterizing stable populations, and application in the African elephant population. J. Wildlife Manage. 37:513–523.

REFERENCES

Francis, R. C., G. A. MacFarlane, A. B. Hallowed, G. L. Swartzman and W. M. Getz, 1984. Status and management of the pacific hake (*Merluccius productus*) resource and fishery of the west coast of the United States and Canada. NWFAC Processed Report 84-118, Natl. Marine Fish. Ser., U. S. Dept. of Commerce.

Gertner, G. Z., and P. J. Dzialowy, 1984. Effects of measurement errors on an individual tree based growth projection system. Can. J. For. Res. 14:311–316.

Getz, W. M., 1979. Optimal harvesting of structured populations. Math. Biosci. 44:269–291.

Getz, W. M., 1980a. The ultimate sustainable yield problem in nonlinear age-structured populations. Math. Biosci. 48:279–292.

Getz, W. M., 1980b. Harvesting models and stock-recruitment curves in fisheries management. In: W. M. Getz (ed.), Mathemathical Modelling in Biology and Ecology. Springer-Verlag, Heidelberg.

Getz, W. M., 1984a. Production models for nonlinear stochastic age-structured fisheries. Math. Biosci. 69:11–30.

Getz, W. M., 1984b. Population dynamics: A resource per capita approach. J. Theor. Biol. 108:623–644.

Getz, W. M., 1985. Optimal and feedback strategies for managing multicohort populations. J. Optimization Theory and Application 46:505–514.

Getz, W. M., 1988. Harvesting discrete nonlinear age and stage structured populations. J. Optimization Theory and Application 57:69–83.

Getz, W. M., and M. O. Bergh, 1988. Quota setting in stochastic fisheries. In: W. S. Wooster (ed.), Biological Objectives and Fishery Management. Springer-Verlag, Berlin.

Getz, W. M., and A. P. Gutierrez, 1982. A perspective in system analysis as applied to crop production and in-

367

sect pest management in agriculture. Ann. Rev. Entomol. 27:447–466.

Getz, W. M., and G. L. Swartzman, 1981. A probability transition matrix model for yield estimation in fisheries. Can. J. Fish. Aq. Sci. 38:847–855.

Getz, W. M., R. C. Francis, and G. L. Swartzman, 1987. Managing variable marine fisheries. Can. J. Fish. Aq. Sci. 44:1370–1375.

Getz, W. M., G. L. Swartzman, and R. C. Francis, 1985. A conceptual model for multispecies, multifleet fisheries. In: M. Mangel (ed.), Proceedings of the R. Yorque Workshop on Resource Management, pp. 49–63. Springer-Verlag, New York.

Ginzburg, L. R., 1986. The theory of population dynamics. I: Back to first principles. J. Theor. Biol. 122:385–399.

Goodwin, G. C., and K. S. Sin, 1984. Adaptive Filtering Prediction and Control. Prentice Hall, Englewood Cliffs, N. J.

Gordon, D. T., 1970. Natural regeneration of white and red fir: the influence of several factors. USDA Forest Serv. Res. Paper PSW-58, Pacific Southwest Forest Exp. Stn., Berkeley, Calif.

Gordon, D. T., 1973. Released advance reproduction of white fir and red fir: Growth, damage, and mortality. USDA Forest Serv. Res. Paper PSW-95, Pacific Southwest Forest Exp. Stn., Berkeley, Calif.

Gordon, D. T., 1979. Successful natural regeneration cuttings in California true fir. USDA Forest Serv. Res. Paper PSW-140, Pacific Southwest Forest Exp. Stn., Berkeley, Calif.

Gordon, H. S., 1954. The economic theory of a common property resource: The fishery. J. Polit. Econ. 62:124—142.

Guckenheimer, J., G. F. Oster, and A. Ipaktchi, 1976. The

dynamics of density dependent population models. J. Math. Biol. 4:101–147.

Gulland, J. A., 1983. Fish Stock Assessment: A Manual of Basic Methods. Wiley, New York.

Gutierrez, A. P., Baumgaertner, and C. G. Summers, 1981. A conceptual model for growth, development and reproduction in the ladybird beetle, *Hippodamia convergens* (Coleoptera: Coccinellidae). Can. Ent. 113:21–33.

Haight, R. G., 1985. A comparison of dynamic and static economic models of uneven-aged stand management. Forest Sci. 31:957–974.

Haight, R. G., 1987. Evaluating the efficiency of even-aged and uneven-aged stand management. Forest Sci. 33:116–134.

Haight, R. G., and W. M. Getz, 1987a. A comparison of stage-structured and single-tree models for projecting forest stands. Natural Resource Modelling 2:279–298.

Haight, R. G., and W. M. Getz, 1987b. Fixed and equilibruim endpoint problems in uneven-aged stand management. Forest Sci. 33:908–931.

Haight, R. G., J. D. Brodie, and D. M. Adams, 1985. Optimizing the sequence of diameter distributions and selection harvests for uneven-aged stand management. Forest Sci. 31:451–462.

Haight, R. G., J. D. Brodie, and W. G. Dahms, 1985. A dynamic programming algorithm for optimization of lodgepole pine management. Forest Sci. 31:321–330.

Hall, D. O., 1983. Financial maturity for even-aged and all-aged stands. Forest Sci. 29:833–836.

Hämäläinen, R. P., A. Haurie, and V. Kaitala, 1984. Bargaining on whales: A differential game with Pareto optimal equilibria. Operations Research Letters 3:5–11.

REFERENCES

Hämäläinen, R. P., A. Haurie, and V. Kaitala, 1985. Equilibrium and threats in a fisheries management game. Optimal Control Applications and Methods 6:315–333.

Hamilton, D. A., 1986. A logistic model of mortality in thinned and unthinned mixed conifer stands of northern Idaho. Forest Sci. 32:989–1000.

Hann, D. W., 1980. Development and evaluation of an even-aged and uneven-aged ponderosa pine Arizona fescue stand simulator. USDA For. Serv. Res. Pap. INT-267, Intermt. Forest and Range Exp. Stn., Ogden, Utah.

Hardie, I. W., J. N. Daberkow, and K. E. McConnell, 1984. A timber harvesting model with variable rotation lengths. Forest Sci. 30:511–523.

Hardin, G., 1968. The tragedy of the commons. Science 162:1243–1247.

Harwood, J., 1978. The effect of management policies on the stability and resilience of British grey seal populations. J. Appl. Ecol. 15:413–421.

Harwood, J., 1981. Managing gray seal populations for optimum stability. In: C. W. Fowler and T. D. Smith (eds.), Dynamics of Large Mammal Populations, pp. 159–172. Wiley, New York.

Hasse, W. D., and A. R. Ek, 1981. A simulated comparison of yields for even-aged versus uneven-aged management of northern hardwood stands. J. Env. Manage. 12:235–246.

Haurie, A., 1982. Stability and optimal exploitation over an infinite time horizon of interacting populations. Optimal Control Applications and Methods 3:241–256.

Hennemuth, R. C., J. E. Palmer, and B. E. Brown, 1980. A statistical description of recruitment in eighteen selected fish stocks. J. Northwest Atl. Fish. Sci. 1:101–111.

REFERENCES

Henry, L., 1976. Population: Analysis and Models. Edward Arnold, London.

Hightower, J. E., and G. D. Grossman, 1985. Comparison of constant effort harvest policies for fish stocks with variable recruitment. Can. J. Fish. Aq. Sci. 42:982–988.

Hightower, J. E., and G. D. Grossman, 1987. Optimal policies for rehabilitation of overexploited stocks. Can. J. Fish. Aq. Sci. 44:803–810.

Hightower, J. E., and W. H. Lenarz, in press. Optimal harvesting policies for the widow rockfish fishery. In: E. A. Vetter and B. E. Megrey (eds.), Mathematical Analysis of Stock Dynamics.

Hobart, W. (ed.), 1985. Pacific Whiting. Marine Fisheries Review 47(2). Scientific Publications Office, NMFS, NOAA, Seattle.

Hochberg, M. E., J. Pickering, and W. M. Getz, 1986. Evaluation of phenology models using field data: Case study for the pea aphid, *Acyrthosiphon pisum*, and the blue alfalfa aphid, *Acyrthosiphon kondoi* (Homoptera: Aphididea). Environ. Ent. 15:227–231.

Hoganson, H. M., and D. W. Rose, 1984. A simulation approach for optimal timber management scheduling. Forest Sci. 30:220–238.

Horwood, J. W., 1982. The variance of population and yield from an age-structured stock, with application to North Sea herring. J. Cons. Int. Explor. Mer. 40:237–244.

Horwood, J. W., 1983. A general linear theory for the variance of yield from fish stocks. Math. Biosci. 64:203–225.

Horwood, J. W., and J. G. Shepherd, 1981. The sensitivity of age-structured populations to environmental variability. Math. Biosci. 57:59–82.

Huang, B., and C. J. Walters, 1983. Cohort analysis and

population dynamics of large yellow croaker in the China sea. N. Am. J. Fish. Manage. 3:295–305.

Hughes, R. D., 1963. Population dynamics of the cabbage aphid *Brevicoryne brassicae* (L.). J. Anim. Ecol. 32:393–426.

Hughes, R. D., and N. Gilbert, 1968. A model of an aphid population—a general statement. J. Anim. Ecol. 37:553–563.

Husch, B., C. I. Miller, and T. W. Beers. 1982. Forest Mensuration, 3d edition. Wiley, New York.

Hyde, W. F., 1976. Economics of national forest timber harvests. J. Forestry 74:823–824.

Intriligator, M. D., 1971. Mathematical Optimization and Economic Theory. Prentice-Hall, Englewood Cliffs, N.J.

Johansson, P. O., and K. G. Löfgren, 1985. The Economics of Forestry and Natural Resources. Basil Blackwell, Oxford.

Johnson, K. N., and H. L. Scheurman, 1977. Techniques for prescribing optimal timber harvest and investment under different objectives. Forest Sci. Monogr. 18.

Johnson, K. N., D. B. Jones, and B. Kent, 1980. A user's guide to the forest planning model (FORPLAN). Land Management Planning, USDA Forest Service, Ft. Collins, Colo.

Kaitala, V., 1986. Game theory models in fisheries management—A survey. In: T. Basar (ed.), Dynamic Games and Applications in Economics, pp. 252–266. Springer-Verlag, Berlin.

Keyfitz, N., 1968. Introduction to the Mathematics of Population. Addison-Wesley, Reading, Mass.

Kimura, D. K., J. W. Balsiger, and D. H. Ito, 1984. Generalized stock recruitment analysis. Can. J. Fish. Aq. Sci. 41:1325–1333.

REFERENCES

Knapp, K. C., 1983. Steady-state solutions to dynamic optimization models with inequality constraints. Land Economics 59:300–304.

Knoebel, B. R., H. E. Burkhart, and D. E. Beck, 1986. A growth and yield model for thinned stands of yellow poplar. Forest Sci. Monogr. 27.

Lefkovitch, L. P., 1965. The study of population growth in organisms grouped by stages. Biometrics 21:1–18.

Lefkovitch, L. P., 1967. A theoretical evaluation of population growth after removing inviduals from some age groups. Bull. Ent. Res. 57:437–445.

Leslie, P. H., 1945. On the use of matrices in certain population mathematics. Biometrika 33:183–212.

Leslie, P. H., 1948. Some further notes on the use of matrices in population mathematics. Biometrika 35:213–245.

Levin, S. A., and C. P. Goodyear, 1980. Analysis of an age-structured fishery model. J. Math. Biol. 9:245–274.

Levin, S. A., and R. M. May, 1976. A note on difference-delay equations. Theor. Pop. Biol. 9:178–187.

Lewis, E. G., 1942. On the generation and growth of population. Sankhya 6:93–96.

Lovejoy, W. S., 1986. Bounds on the optimal age-at-first-capture for stochastic, age-structured fisheries. Can. J. Fish. Aq. Sci. 41:101–107.

Ludwig, D., and C. J. Walters, 1981. Measurement errors and uncertainty in parameter estimates for stock recruitment. Can. J. Fish. Aq. Sci. 38:711–720.

Luenberger, D. G., 1979. Introduction to Dynamic Systems: Theory, Models, and Applications. Wiley, New York.

Lyon, K. S., and R. A. Sedjo, 1983. An optimal control theory model to estimate the regional long-term supply of timber. Forest Sci. 29:798–812.

Lyon, K. S., and R. A. Sedjo, 1986. Binary-search SPOC:

An optimal control theory version of ECHO. Forest Sci. 32:576–584.

Mangel, M., 1985. Decision and Control in Uncertain Resource Systems. Academic Press, Orlando, Florida.

May, R. M., and G. F. Oster, 1976. Bifurcations and dynamic complexity in simple ecological models. Am. Natur. 110:573–590.

May, R. M., J. R. Beddington, J. W. Horwood, and J. G. Shepherd, 1978. Exploiting natrual populations in an uncertain world. Math. Biosci. 42:219–252.

May, R. M., G. R. Conway, M. P. Hassell, and T.R.E. Southwood, 1974. Time delays, density dependence, and single species oscillations. J. Anim. Ecol. 43:747–770.

Mendelssohn, R., 1978. Optimal harvesting strategies for stochastic single species, multiage class models. Math. Biosci. 41:159–174.

Mercer, M. C. (ed.), 1982. Multispecies approach to fisheries management advice. Can. Spec. Publ. Fish. Aq. Sci. 59.

Michie, B. R., 1985. Uneven-aged stand management and the value of forest land. Forest Sci. 31:116–121.

Michie, B. R., and J. Buongiorno, 1984. Estimation of a matrix model of forest growth from remeasured permanent plots. Forest Ecol. and Manage. 8:127–135.

Murawski, S. A., 1984. Mixed-species yield-per-recruitment analysis accounting for technological interactions. Can. J. Fish. Aq. Sci. 41:897–916.

Murawski, S. A., A. M. Lange, M. P. Sissenwine, and R. K. Mayo, 1983. Definition and analysis of multispecies otter-trawl fisheries off the northeast coast of the United States. J. Cons. Int. Explor. Mer. 41:13–27.

Nautiyal, J. C., and P. H. Pearse, 1967. Optimizing the conversion to sustained yield—a programming solution. Forest Sci. 13:131–139.

REFERENCES

Navon, D. I., 1971. Timber RAM: A long-range planning method for commercial timber lands under multiple-use management. USDA Forest Serv. Res. Paper PSW-70, Pacific Southwest For. and Range Exp. Stn., Berkeley, Calif.

Nazareth, L., 1980. A land management model using Dantzig-Wolfe decomposition. Manage. Sci. 26:510–523.

Nisbet, R. M., and W.S.C. Gurney, 1982. Modelling Fluctuating Populations. Wiley, New York.

Odum, H. T., 1983. Systems Ecology. Wiley, New York.

Parry, B. T., H. J. Vaux, and N. Dennis, 1983. Changing conceptions of sustained yield policy on the National Forests. J. Forestry 81:150–154.

Patten, B. C., 1971. Systems Analysis and Simulation in Ecology, Volume 1. Academic Press, New York.

Patten, B. C., 1972. Systems Analysis and Simulation in Ecology, Volume 2. Academic Press, New York.

Peterman, R. M., 1981. Form of random variation in salmon smolt-to-adult relations and its influence on production estimates. Can. J. Fish. Aq. Sci. 38:1113–1119.

Pikitch, E. K., 1987. Use of a mixed species yield-per-recruit model to explore the consequences of various management policies for the Oregon flatfish fishery. Can. J. Fish. Aq. Sci. 44(Suppl. 2):349–359.

Plant, R. E., 1986. The sterile insect technique: A theoretical perspective. In: M. Mangel (ed.), Systems Analysis in Fruitfly Management. Springer-Verlag, Heidelberg.

Pollard, J. H., 1966. On the use of the direct matrix product in analysing certain stochastic population models. Biometrika 53:397–415.

Pollard, J. H., 1973. Mathematical Models for the Growth of Human Populations. Cambridge University Press, Cambridge.

REFERENCES

Pope, J. G., 1972. An investigation of the accuracy of virtual population analysis using cohort analysis. Res. Bull. ICNAF 9:65–74.

Pope, J. G., 1979. Stock assessment in multispecies fisheries with special reference to the trawl fishery in the Gulf of Thailand. FAO South China Sea Fisheries Development Program. SCS/DEV/79/19.

Pope, J. G., and J. G. Shepherd, 1982. A simple method for the consistent interpretation of catch-at-age data. J. Cons. Int. Explor. Mer. 40:176–184.

Pope, J. G., and J. G. Shepherd, 1985. A comparison of the performance of various methods for tuning VPA using effort data. J. Cons. Int. Explor. Mer. 42:129–151.

Rao, C. R., 1973. Linear Statistical Inference and Its Application, 2d edition. Wiley, New York.

Reed, W. J., 1979. Optimal escapement levels in stochastic and deterministic harvesting models. J. Environ. Econ. Manage. 6:350–363.

Reed, W. J., 1980. Optimum age-specific harvesting in a nonlinear population model. Biometrics 36:579–593.

Reed, W. J., 1983. Recruitment variability and age-structure in harvested populations. Math. Biosci. 65:239–268.

Reed, W. J., 1986. Optimal harvesting models in forest management—a survey. Natural Resource Modeling 1:55–79.

Reed, W. J., and D. Errico, 1986. Optimal harvest scheduling at the forest level in the presence of the risk of fire. Can. J. For. Res. 16:266–278.

Rexstad, E. A., and E. K. Pikitch, 1986. Stomach contents and food consumption estimates of pacific hake, *Merluccius productus*. Fishery Bulletin 84:947–956.

Reynolds, M. R., 1984. Estimating the error in model predictions. Forest Sci. 30:454–469.

REFERENCES

Ricker, W. E., 1954. Stock and recruitment. J. Fish. Res. Board Can. 11:559–623.

Rideout, D., 1985. Managerial finance for silvicultural systems. Can. J. For. Res. 15:163–166.

Riitters, K., J. D. Brodie, and D. W. Hann, 1982. Dynamic programming for optimization of timber production and grazing in Ponderosa pine. Forest Sci. 28:517–526.

Roise, J. P., 1986. An approach to optimizing residual diameter class distributions when thinning even-aged stands. Forest Sci. 32:871–881.

Rorres, C., 1978. A linear programming approach to the optimal sustainable harvesting of a forest. J. Environ. Manage. 6:245–254.

Rorres, C., and W. Fair, 1975. Optimal harvesting policy for an age-specific population. Math. Biosci. 24:31–47.

Rothschild, B. J., 1986. Dyamics of Marine Fish Populations. Harvard University Press, Cambridge, Mass.

Rubenstein, R. Y., 1981. Simulation and the Monte Carlo Method. Wiley, New York.

Ruppert, D., R. L. Reish, R. B. Deriso, and R. J. Carroll, 1984. Optimization using stochastic approximation and Monte Carlo simulation (with application to harvesting Atlantic menhaden). Biometrics 40:535–546.

Ruppert, D., R. L. Reish, R. B. Deriso, and R. J. Carroll, 1985. A stochastic population model for managing the Atlantic menhaden (*Brevoortia tyrannus*) fishery and assessing managerial risks. Can. J. Fish. Aq. Sci. 42:1371–1379.

Schnute, J., 1985. A general theory for analysis of catch and effort data. Can. J. Fish. Aq. Sci. 42:414–429.

Schnute, J., 1987. A general fishery model for a size-structured fish populations. Can. J. Fish. Aq. Sci. 44:924–940.

REFERENCES

Schnute, J., and D. Fournier, 1980. A new approach to length frequency analysis: Growth structure. Can. J. Fish. Aq. Sci. 37:1337–1351.

Shepherd, J. G., and M. D. Nicholson, 1986. Use and abuse of multiplicative models in the analysis of fish catch-at-age data. The Statistician 35:221–227.

Shoukry, A., and M. Hafez, 1979. Studies on the biology of the Mediterranean fruit fly, *Ceratitis capitata*. Entomol. Exp. Appl. 26:33–39.

Sinko, J. W., and W. Streifer, 1967. A new model for age-structure of a population. Ecology 48:910–918.

Smith, D. M., 1962. The practice of silviculture. Wiley, New York.

Swartzman, G. L. and S. P. Kaluzny, 1987. Ecological Simulation Primer. Macmillan, New York.

Swartzman, G. L., W. M. Getz, R. C. Francis, R. Haar, and K. Rose, 1983. A management analysis of the Pacific whiting fishery using an age-structured stochastic recruitment model. Can. J. Fish. Aq. Sci. 40:524–529.

Swartzman, G. L., W. M. Getz, and R. C. Francis, 1987. Bi-national management of Pacific hake: A stochastic modeling approach. Canad. J. Fish. Aq. Sci. 44:1053–1063.

Sykes, Z. M., 1969. On discrete stable population theory. Biometrics 25:285–293.

TenEyck, N., and R. Demory, 1975. Utilization of flatfish caught by Oregon trawlers in 1974. Oregon Dept. Fish. Wildlife Information Rep. 75-3, ODFW, Newport, Oregon.

Travis, C. C., W. M. Post, D. L. DeAngelis, and J. Perkowski, 1980. Analysis of compensatory Leslie matrix models for competing species. Theor. Pop. Biol. 18:16–30.

Tuljapurkar, S., 1987. Cycles in nonlinear age-structured models. I: Renewal equations. Theor. Pop. Biol. 32:26–41.

378

Tuljapurkar, S., in press. Cycles in nonlinear age-structured models. II: McKendrick-von Foerster equations. Theor. Pop. Biol.

USDA Forest Service, 1981. Draft environmental impact statement 05-15-81-02. Sierra National Forest, Fresno, Calif.

USDA Forest Service, 1983. Silvicultural systems for the major forest types of the Unites States. USDA Forest Service Agricultural Handbook No. 445.

Usher, M. B., 1966. A matrix approach to the management of renewable resources, with special reference to selection forests. J. Appl. Ecol. 3:355–367.

Usher, M. B., 1969a. A matrix model for forest management. Biometrics 25:309–315.

Usher, M. B., 1969b. A matrix approach to the management of renewable resources, with special reference to selection forests—two extensions. J. Appl. Ecol. 6:347–348.

Usher, M. B., 1976. Extensions to models, used in renewable resource management, which incorporate an arbitrary structure. J. Environ. Manage. 4:123–140.

Van Deusen, P. C., and G. S. Biging, 1985. STAG user's guide: A stand generator for mixed species stands. Res. Note #11, Northern Calif. For. Yield Coop., Dept. of For. and Res. Mgmt., Univ. of Calif., Berkeley, Calif.

Verhulst, P. F., 1838. Notice sur la loi que la population suit dans son accroissement. Corresp. Math. Phys. 10:113–126.

Waggener, T. R., 1977. Community stability as a forest management objective. J. Forestry 75:710–714.

Walters, C. J., 1986. Adaptive Management of Renewable Resources. Macmillan, New York.

Walters, C. J., and R. Hilborn, 1976. Adaptive control of fishing systems. J. Fish. Res. Board Can. 33:145–159.

REFERENCES

Walters, C. J., and D. Ludwig, 1981. Effects of measurement errors on the assessment of stock recruitment relationships. Can. J. Fish. Aq. Sci. 38:704–710.

Wensel, L. C., and J. R. Koehler, 1985. A tree growth projection system for northern California mixed coniferous forests. Res. Note #12, Northern Calif. For. Yield Coop., Dept. of For. and Res. Mgmt., Univ. of Calif., Berkeley.

Wu, L. S.-Y., and D. B. Botkin, 1980. Of elephants and men: A discrete stochastic model for long-lived species with complex life histories. Am. Natur. 116:831–849.

Wykoff, W. R., 1985. Introduction to the PROGNOSIS model version 5.0. In: Growth and Yield and Other Mensurational Tricks: A Regional Technical Conference. USDA For. Serv. Gen. Tech. Rep. INT-93, pp. 44–52. Intermt. For. and Range Exp. Stn., Ogden, Utah.

Wykoff, W. R., N. R. Crookston, and A. R. Stage, 1982. User's guide to the stand prognosis model. USDA For. Serv. Gen. Tech. Rep. INT-133, Intermt. For. and Range Exp. Stn., Ogden, Utah.

Index

INDEX

Weeks, D., 356
Wensel, L. C., 240, 242, 243, 244, 246
white fir (*Abies concolor*): managed stand projections, 255–58; unmanaged stand projections, 250–54. *See also* true fir
whiting (Pacific), 168, 177, 188, 196, 210–20; distribution, 210–11, 212; equilibrium strategies, 218; parameters, 215
whole-stand models. *See* stand growth models, whole-stand
Wu, L. S.-Y., 328–31, 356
Wykoff, W. R., 240, 246, 252, 253

yield, 82, 92, 96–97, 140; aggregated, 193; biomass, 140, 146; cacometric, 149; concave utility, 177; costs, 97, 297; effort curves, 146–48, 152, 195; equilibrium, 147; eumetric, 149, 154; flow con-

straints, 189, 311–13; general policy, 177; isopleths, 142, 148; log criterion, 174–76, 186–88; long-term average, 203; maximum-log, 163; minimum deviation, 164, 165; minimum-stock-deviation, 163, 164; numerical results, 152, 154, 160, 161, 162, 176, 183, 187, 191, 193, 208, 217, 218, 292, 301, 303; optimal, 92, 148–50; per recruit, 82, 141, 146; per unit effort, 149; planning horizon, 93; pupae, 344; quotas, 173; revenue maximization, 97; sensitivity, 178; short vs. long term, 92–93; suboptimal, 150; surplus model, 136; trade-off, 92–93; two-player, 214; value at 0.1 slope, 150, 164; variability, 111, 176, 178, 223; variance, 167. *See also* discounted rent; harvesting; rent; sustainable yield

MONOGRAPHS IN POPULATION BIOLOGY

EDITED BY ROBERT M. MAY

(continued)